Carbon Dharma:
The Occupation of Butterflies

Using the metaphor of metamorphosis, Carbon Dharma calls for our occupation of the Earth as Butterflies, to undo the damage done by the human species in its present Caterpillar stage of existence. It diagnoses the reasons for our Caterpillar stage of existence as the misinterpretation of the fundamental principles that underlie our democracy and our industrial civilization.

The book is intended for the youth of this world who are facing some of the gravest challenges ever faced by any generation of human beings. It is also intended for all those who love the youth of this world, for the youth cannot solve these challenges on their own while their parents, grandparents, uncles and aunts continue to pile on more grave challenges for them to solve.

While drawing upon the ancient Hindu concept of Dharma, or "right action," the book weaves illustrative stories from the author's life and leads up to a global call to action, action of a very specific, focused kind. Rather than listing hundreds of "change-light-bulb" type actions that a lot of us have been doing disjointly, but somewhat ineffectively, it lists three specific actions that we can begin to do concertedly to make a difference. While changing the world is about changing ourselves, effecting social change requires such concerted action.

A Climate Healers Publication

ISBN-13: 978-1467928458
ISBN-10: 1467928453

Email: info@climatehealers.org
Publisher Website: http://www.climatehealers.org
Book Website: http://www.carbondharma.org

Front Cover Photo: Courtesy, the Karech Village Forest Protection Committee of Karech village, Udaipur District, Rajasthan, India, and the Foundation for Ecological Security, Anand, Gujarat, India.

Preface: Courtesy Brian D. McLaren
All Illustrations: Courtesy Niharika Desiraju.

Carbon Dharma:

The Occupation of Butterflies

Sailesh Rao

with Preface by Brian McLaren
and Illustrations by Niharika Desiraju

A Climate Healers Publication

All proceeds from the sale of this book goes to support the 501(c)3 non-profit, Climate Healers.

Sailesh Rao is the Executive Director of the US 501(c)3 non-profit corporation, Climate Healers. An Electrical Engineer by training with a Ph.D., from Stanford University in Stanford, CA, Sailesh switched careers and became deeply immersed in the various environmental crises facing humanity after some life changing events and after he watched Vice President Al Gore's slide show on TV. Carbon Dharma is the result of his findings over the years.

Preface is by Brian D. McLaren, an author, speaker, activist and public theologian. More on Brian at http://www.brianmclaren.net

All illustrations are by Niharika Desiraju of Danville, CA, USA, who drew them in the summer between her fifth and sixth grades.

Table of Contents

For Sushil, Akhil and the Miglets of the world.

"If there was no change, there would be no Butterflies," sign at an Occupy San Francisco Rally.

Preface

In biology, there's a phenomenon called parallel evolution. Two different species found on two different continents from two distant branches of the evolutionary tree evolve similar characteristics to survive and thrive.

How does this happen?

Similar conditions, biologists tell us, mean that similar characteristics provide a survival advantage.

Something similar is happening in the world of religion. Sailesh Rao is a Hindu. I am a Christian. Some might say we are different religious species from different continents from distant branches on the family tree of human religion. But as I've gotten to know Sailesh, as I've learned of his beautiful work among the poor of India, and even more as I've read Carbon Dharma, I feel I've found a spiritual brother, a kindred spirit.

Coming from such different backgrounds, we've reached remarkably similar (and where they differ, complementary) conclusions about what's wrong in our world and what needs to be done to make it right.

My version of the journey began in 2006. After leaving academia in the 1980s to become a pastor in a nondenominational Christian church, I started writing books. In 2005, I wrote a book on the essential message of Jesus, which I discovered to be (this is pretty obvious, but sometimes religious people show amazing creativity in missing the point) "the good news of the kingdom of God." That pregnant phrase, I had become convinced, was not about where we go after we die, as I had been taught: it was about how we live before we die. It was a call to reconciliation - with God, with ourselves, with our neighbors and even our enemies, and with nonhuman creation as well.

When I left pastoral work to write full-time in 2006, I knew what I wanted my next research and writing project to be. First, I wanted to pursue a question that had been nagging me for over twenty years, simmering on the back burner, so to speak: what are the world's biggest problems? In other words, which problems are the diseases beneath the symptoms ... which problems the game-changers (or enders) ... which problems, if they aren't addressed, could destroy us first or hurt us the worst?

Then, I wanted to take what I had been discovering about Jesus and his message and apply it to those questions. The result was a book called Everything Must Change: Where the World's Biggest Problems and Jesus' Good News Collide. Writing Everything Must Change certainly changed me. I'll never recover from that research and reflection.

Meanwhile, as you'll read in the chapters that follow, a Hindu engineer from India was going through his own evolutionary process. In the end, we both arrived at similar conclusions.

1. Our world is in trouble, big trouble.
2. The big trouble relates to our relationship with the environment.
3. We can't make the changes we need without a deep spiritual change.
4. Our respective religious traditions offer powerful and profound resources to help us experience that change so that we are converted from being part of the global problem to being part of the global solution.

But it wasn't only the similarities between our work that intrigued me as I read Carbon Dharma. I was even more intrigued to gain greater insight into Hinduism and how Hindu perspectives, rooted in the Hindu scriptures, can enrich our pursuit of a better, more sustainable, even regenerative way of life.

There's a section in the pages to follow where Sailesh tells about his struggle as a boy coming of age to deal with the stories of Hindu scripture. Must he take them literally? Are they only true or valid or valuable if they are literal? Or can they be actual without being factual? I found a huge smile spreading across my face as I recalled my own parallel structures as a boy caught between scientific evidence for a universe 14 billion years old and a biblical story that unfolded some 6000 years ago.

So you won't be surprised how I, as a committed Christian, feel so enthusiastic about a book that provides Hindu perspective on the environment and our place in it. However distant our past evolution has been, one thing is certain: our futures are interwoven and the challenges we face require us all to bring all our best resources to the table - scientific, yes, but also spiritual.

The importance of spiritual resources became all the more clear to me about a year after my book on global crises was published. The advisor to a prime minister contacted me and asked to speak with me. The head of state for whom he worked was a leader in the international community arguing for serious and sustained action on behalf of the environment in general and climate in particular. The prime minister, he explained, had been impressed with the book and gave copies to his whole senior staff, including this advisor.

"The reason I wanted to meet with you," he explained, "was to tell you that I've lost hope about political solutions successfully dealing with this crisis. As your country makes clear, political processes reward politicians who can exploit short-term, hot-button issues to get elected and re-elected. Political processes punish political leaders like my prime minister who try to deal with long-term issues like climate and the environment."

He went on to explain the conclusion he had reached: the only way needed change would come would be through a world-wide, grass-roots spiritual movement that focused on changing hearts first, and then focused on changing behavior. Then we can move on to changing policies - and politicians.

May Carbon Dharma contribute to the growth of exactly that kind of world-wide, grass-roots, multi-religious spiritual movement.
Brian D. McLaren

Prologue

"Thought is the blossom; language the bud; action the fruit behind it" - Ralph Waldo Emerson.

It was just after I gave a talk at the NASA Jet Propulsion Laboratories (JPL) in Pasadena, California, in November 2008 that my dear friend, Gani, urged me to write this book. Gani is Dr. Gani Ganapathi, the rocket scientist from among all my classmates from India and he worked on the heat rejection system for the incredibly successful Mars Exploration Rovers at JPL. His assessment that I had something unique to contribute to the discourse was eye-opening to me. And from time to time, Gani kept nudging me to write the book and so did a number of my other friends, Joe Murray while helping me with my non-profit work, Manju Seal whose daughter Niharika did the wonderful illustrations for the book, and my son Sushil, among others. But nothing concrete happened until I held our granddaughter, Kimaya, in my arms two years later.

Kimaya was born to our son Akhil and the lovely Roxy Chappell on November 19, 2010, in Phoenix, Arizona. Akhil is South Asian, while Roxy is half Native American and half African American, with the result that Kimaya, Sanskrit for "Divine" or "Heavenly," with the direct lineage of at least three continents, was the personification of all humanity in my eyes.

Now I had to write. I had to write down whatever I've understood so far, knowing that it will make at least a small difference in the kind of world that she inherits.

It has been exactly thirty years since I landed in America with an admission to a graduate program at the State University of New York, a teaching assistantship to help pay my way through graduate school, and $100 in my pocket. And what a ride it has been! I got my Masters at the State University of New York at Stony Brook in Long Island, NY, went on to do a Ph.D. at Stanford University in

California, and met and married my beautiful wife, Jaine, while at Stanford. We moved to New Jersey, had two absolutely adorable children, while I carved out an amazing career in technology, reaching the pinnacle of my professional success when the company that had acquired ours, Level One Communications, was bought by Intel Corporation in 1999 for 2.2 billion dollars. We became millionaires and I was on top of the world! At that point, I thought that we had achieved our goal of ensuring security and wealth for our two children and for our old age!

Then our sons turned teenagers and the troubles began. There were episodes of drunken behavior that I attributed to the fact that we had irresponsible neighbors who fed our underage kids alcohol in their homes. I tried to ban our older child, Sushil, from associating with the neighbors' children, but that didn't work. When Sushil, a really bright kid, got admitted to early college right after his junior year in high school, we readily agreed so that he could get away from our neighborhood. Two years and a whole slew of bad grades later, Sushil entered a rehab facility in Arizona after getting arrested for possession of marijuana while driving in upstate New York. Sushil has now been sober for over six years and lives with his beautiful spouse, Michelle, just minutes from us. But when that arrest happened in 2005, I knew that something was really wrong with the life that I was leading, that I was being whacked on the head and told to wake up. Later that year, I saw former Vice President Al Gore's slide show on TV and it was then that I began to suspect that something was fundamentally wrong with the life that we're all leading as human beings.

But my initial instinct was just like my reaction to our neighbor's liberal alcohol policy, to try and see if we can solve the problem while maintaining what Mr. Gore calls "civilization as we know it." It was only when our younger son, Akhil, while a student at Lehigh University, attempted to commit suicide in September 2007, that I felt the urge to truly understand what was at the heart of it all. There were questions aplenty. How could this incredibly wonderful and loving child have grown to become so miserable that he lost the will to continue living? What did I do wrong? What did we all do wrong? And most importantly, why did this happen?

During my years at Stanford, I had a very wise advisor, Prof. Thomas Kailath, who told me that the holder of the degree of Doctor of Philosophy from Stanford University should have acquired the ability to understand any problem in any field, document it and propose credible solutions. With that in mind, I set about attempting to understand and document the system level problem that is our modern industrial civilization. This book is an account of my findings over the past four years and I hope that it meets the exacting standards that Prof. Kailath set for all his students.

The great 20th century philosopher, Anthony DeMello wrote in his book, Awareness[1], "The trouble with people is that they are busy fixing things that they don't even understand... It never strikes us that things don't need to be fixed... They need to be understood. If we understood them, they'd change. Do you want to change the world? How about beginning with yourself?.. Through observation, through understanding, with no interference or judgement on your part.

What you judge, you cannot understand... Observe without a desire to change what is. Because, if you desire to change what is into what you think should be, you no longer understand... The day you attain a posture like that, you will experience a miracle. You will change - effortlessly, correctly. Change will happen, you will not have to bring it about. As the light of awareness settles upon the darkness, whatever is evil will disappear. Whatever is good will be fostered."

With awareness comes understanding and with understanding comes change and action. Here's wishing that such awareness will spread throughout the world and that this book nudges the reader towards this worthy goal.

The book is intended for the youth of this world who are facing some of the gravest challenges ever faced by any generation of human beings. It is also intended for all those who love the youth of this world, for they cannot solve these challenges on their own while their parents, grandparents, uncles and aunts continue to pile on more grave challenges for them to solve.

The book draws upon ancient Hindu texts for that's what I've become most familiar with and it is written from the perspective of a South Asian immigrant to America, for that's who I am. But this is just a reflection of my limitations as an individual and author. The book is intended to be a global call to action, action of a very specific, focused kind. Rather than listing hundreds of "change-light-bulb" type actions that a lot of us have been doing disjointly, but somewhat ineffectively, it lists three specific actions that we can begin to do concertedly to make a difference. While changing the world is about changing ourselves, effecting social change requires such concerted action. Mahatma Gandhi is often quoted as saying, "Be the change you want to see in the world," but a recent investigation by Brian Morton in the New York Times[2] showed that this quote is a bumper-sticker style corruption of Gandhi's words. The closest he could find was Gandhi saying, "If we could change ourselves, the tendencies in the world would also change. As a man changes his own nature, so does the attitude of the world change towards him. ... We need not wait to see what others do".

Personal and social transformation go hand in hand. As Brian Morton explained in the article, "For Gandhi, the struggle to bring about a better world involved not only stringent self-denial and rigorous adherence to the philosophy of nonviolence; it also involved a steady awareness that one person, alone, can't change anything, an awareness that unjust authority can be overturned only by great numbers of people working together with discipline and persistence".

Here's wishing that we will work together with such discipline and thereby, persevere.

Sailesh Rao
Danville, CA, Oct. 2011.

Carbon Dharma

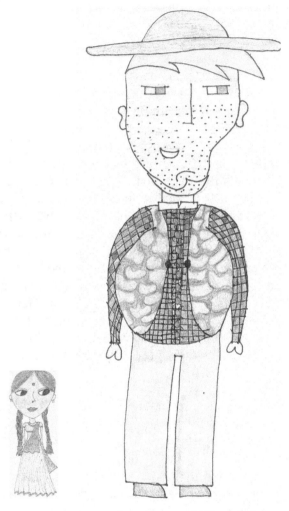

Fig. 1. *Relative impacts of the bottom 20% and the top 20% on planetary resources in the Caterpillar culture.*

1. The Caterpillar and The Butterfly

"Knowledge is proud that it knows so much. Wisdom is humble that it knows no more " - William Cowper.

The Caterpillar wriggles out and proceeds to eat the nutritious shell of the egg that its mother had laid on the underside of the leaf. Then it eats the leaf that the egg was clinging to. And it continues on for the next few weeks eating all the leaves that it encounters. The Caterpillar is a voracious consumer, a veritable eating machine, a crawling stomach.

Once fully satiated and grown to its adult size, the Caterpillar attaches to the underside of a twig and turns into a Pupa. A couple of weeks pass by and the Butterfly emerges. The Butterfly spreads its wings and flies about sipping nectar from the flowers. It is a very discriminating, light consumer. As it sips nectar, the Butterfly pollinates the flowers helping to regenerate Life.

The Butterfly undoes the destruction that the Caterpillar wrought and then some. This is why the Butterfly is a net asset to Life on Earth despite the destructive nature of its Caterpillar stage.

As a species, Homo Sapiens, Latin for "Wise Man," is most definitely in its Caterpillar stage of development. It is a voracious consumer, a waste-producing eating machine that has munched through most of the complex Life in the ocean and half the forests on land.

As a species, I believe that it is well past due for Homo Sapiens to grow into its pupa stage, emerge as a Butterfly and begin undoing the destruction wrought earlier.

For Life only supports that which supports Life. A dominant species that voraciously consumes Life and never regenerates it will eventually consume all Life and consequently, itself.

The question is, how do we achieve such a metamorphosis? How can we all live our lives as Butterflies, as a matter of course?

In his excellent book, "Blessed Unrest[1]," Paul Hawken showed that humanity has already begun its transformation from its Caterpillar stage to its Butterfly stage. Literally, millions of Non-Governmental Organizations (NGOs) dedicated to healing the environment or dedicated to healing the divisions within human society have sprung up around the world as an immune response to the ravages of our modern industrial civilization.

A couple from New Jersey, Pam and Anil Malhotra, migrate to the Kodagu district in Karnataka, India, in 1991 and purchase 55 acres of a coffee plantation in private lands surrounded by the Brahmagiri, Nagarhole and Bandipur National Forests. Over the years, they purchase neighboring private land holdings and slowly accumulate over 300 acres of land, which they simply donate back to Nature by leaving the land alone. Their neighbors think that they are crazy to be purchasing coffee plantations without the intention of growing any coffee. But over a twenty year period, Nature blooms back on their private reserve with JackFruit trees, elephant herds, birds and insects thriving as the forest regenerates. My sister, Sudha, and I stayed in a cottage on their Save Animals Initiative (SAI) sanctuary[2] and the din from the call of the crickets at night still ring in my ears. Students from American and European universities now make regular visits to the SAI Sanctuary to study the incredible biodiversity that once prevailed throughout the Western Ghats of India.

Pam and Anil Malhotra are truly human Butterflies.

Another example of human Butterflies at work can be seen in the protected common lands throughout the forest villages of India. In conjunction with the NGO, the Foundation for Ecological Security[3] (FES), the Karech Village Forest Protection Committee in the Udaipur district of Rajasthan set aside 250 acres of common land in 2002 by enclosing it with a stone fence to prevent livestock from grazing in there. The villagers also agree not to take any biomass for cooking fuel from within the protected land. Four years later, in 2006, the protected land becomes lush green while the land outside the stone fence remains barren, with new growth eaten by livestock and old growth cut down for firewood. I visited this protected land in December of 2008, saw the contrast across the stone fence and turned vegan as the realization dawned on me that my consumption of dairy products was responsible for causing the forests to disappear. Since then, I haven't deliberately consumed or purchased a single item tainted with animal matter.

In contrast with Butterflies, human Caterpillars do the opposite transformation. Caterpillars take a pristine forest, cut it down in order to either extract some mineral from under it or to convert it to pasture land or agricultural land, leaving pollution and waste behind, typically ruining the livelihoods of native peoples within the forest. An example of Caterpillars at work can be found in the Canadian Tar Sands of Athabasca in Alberta, Canada[4], where a pristine Boreal forest the size of the state of Delaware was cut down in just the past five years and converted into polluted wasteland in order to extract the thick bituminous oil from under the forest floor. The First Nations people, who used to live sustainably within the forest, have found their livelihoods disrupted as their fresh water sources are polluted. Perhaps, they will have to seek employment driving the bulldozers and the mining equipment of the Tar Sands oil extraction operations in order to subsist. Oil companies have leased a Florida-sized area of Boreal forest lands in Canada with the intention of meting out the same treatment to all that land in the coming years. This is how Life is devoured and native cultures destroyed by human Caterpillars at work.

Butterflies heal, while Caterpillars destroy.

At present, despite Paul Hawken's optimistic projections and the Butterfly examples cited above, the Caterpillars are far more numerous and more potent than the Butterflies and as a result, the destruction on Earth far outpaces the healing. A lot of the destruction caused by Caterpillars is permanent for all practical purposes. For example, it is difficult to imagine the Boreal forests regenerating any time soon from the vast clear cutting and the cesspool of chemical pollution resulting from the Tar Sands oil extraction process. But Nature has always surprised us with her resilience. Even in Chernobyl, Ukraine, 25 years of human avoidance of its radioactive environs has resulted in the regeneration of an ecosystem with wolves using abandoned buildings for their lairs and eagles nesting in vacant penthouses[5]. However, in the long run, the wolves and eagles are going to suffer the consequences of the silent killers that humans unleashed in Chernobyl, the radioactive waste.

1.1 The Population Conundrum

Whether we behave like Caterpillars or whether we behave like Butterflies depends on the stories that we have bought into, which then rule our actions. It has now become commonly accepted lore that human beings have no choice but to be Caterpillars and from this, it logically follows that the sheer number of human beings on the planet is the primary reason for the destruction that we're inflicting on the planet. That is, we are led to believe that all our environmental problems are mainly due to human "overpopulation".

In 1968, Paul and Anne Ehrlich wrote "The Population Bomb[6]," boldly predicting widespread famines in the world during the 1970s and 1980s and recommending drastic measures for curtailing human population across the world. The Ehrlichs' book propelled the population debate into the mainstream and human population growth has remained a popular bugbear for societal ills in some intellectual circles. With the advent of the industrial revolution, human life expectancy has steadily increased with declining death rates and human population has ballooned. In 1800, human population reached 1 billion; it increased to 1.6 billion by 1900 and 6.1 billion by 2000, marking a six-fold increase in 200 years. Human population is currently at 7 billion and it is projected to top out around 10 billion on the high end by 2100 according to UN projections[7].

Of course, such projections assume that the past is prologue.

While human population increased by a factor of 6 from 1800 to 2000, the inconvenient truth for the population "bombers" is that human consumption of world resources increased by a factor of 64 within the same time frame. Vast numbers of humans are living like the nobility of the past, eating exotic foods from all over the world and traveling vast distances on a frequent basis. Such indulgences use up resources. World consumption, in constant dollars, increased from $400B in 1800 to $1.6T in 1900 to over $25T in 2000 fueling the meteoric rise in human impact on the planet[8].

As human consumption has grown 64-fold in 200 years, the inequity in consumption between the top and the bottom has also grown, becoming increasingly stark in recent times. The top 20% of humans were responsible for 83% of world consumption circa the year 2000, while the bottom 20% were responsible for just 1.3%[9]. This is a 64-fold difference in impact or footprint at the two ends of the social stratum. Most of the increase in human population is occurring in the bottom 20% of the economic pyramid, with about 80 million additional human beings added to the world population each year, while most of the increase in consumption is occurring at the top.

This sets up an Alphonse-Gaston act between the top and the bottom economic tiers of society with both sides talking past each other as they view the world from their narrow perspectives, while doing nothing about the situation. The people in the top 20% see an overpopulation problem as they are concerned that all these additional people would eventually want to live like them and they know that there certainly aren't enough resources on the planet to accommodate this shift. Furthermore, the people in the top 20% also want to improve their state of affluence constantly, expecting a 3% increase in consumption on an annual basis and firing their political leaders when they are unable to achieve such economic growth. The people in the bottom 20% are aggrieved that every child in the top tier has the same impact as 64 children in the bottom tier and yet the additional children they bring into the world in the bottom tier in a quest for survival and security are somehow construed as a burden on society. After all, the top 1.3 billion people on the planet look like 83 billion people from the perspective of the bottom 20% of the pyramid. And from the perspective of the bottom 20%, those 83 billion people at the top ought to be ashamed when they tell the people below them that there are too many of them.

Our current economic structures exacerbate such disparities. For instance, productivity improvement is supposed to be the constant quest of a technological society, but lately, this has become a euphemism for cutting people out of employment or dumbing them down. Coal mines that used to require thousands of laborers can now be operated with a few people and giant earth moving equipment through the process of mountaintop removal mining. Factories that used to require thousands of people to run can now be operated with automated processes manned by just dozens of people. Such productivity gains improve profit margins for the corporations, but reduce employment for human beings. The only way for people to find employment is for the economy to start making even more stuff and to start peddling it to even more numerous and voracious consumers. It is consumption, the fulfillment of desires, that primarily drives economic growth in the current paradigm. There are even serious books entitled, "Why People Buy Things They Don't Need[10]," to teach marketing professionals how to trigger such desires among their potential customers. American homes have almost tripled in size since the 1950s and yet, the American appetite for "stuff" is such that there is a $20B a year industry[11] to store overflow "stuff" for Americans. And yet, the growth paradigm of our economic well-being requires people to buy even more "stuff." This is why Vice President Joe Biden wrote an Op Ed[12] in the New York Times urging the Chinese and other people in developing countries to stop saving and start consuming in order to drive world economic growth. More than citizens or human beings, people have become consumers, whom President Herbert Hoover characterized as[13] "constantly moving happiness machines, machines which have become the key to economic progress."

But this standard formula for economic progress is running out of steam. And it is up to us who are alive today to recognize that and make a shift in the way we live. Specifically, I consider the 10-30 year olds alive today as the Most Important Generation that ever Lived on Earth (MIGLE) for they need to become Butterflies. Or choose not to change and face intense suffering and possibly, a major mass extinction event in their lifetimes. The members of this generation are Miglets. Our two children are Miglets. At their age, Miglets are prone to be idealistic and are much more likely to lead the way towards such fundamental behavioral shifts.

At present, many Miglets have done all the things that their elders wanted of them, studied hard, got a good education, played by the rules or more likely, never got caught breaking rules, but they haven't found jobs in the present economic environment in America. I know one Miglet who was working at a meat packing plant at minimum wage after a BS in Chemical Engineering. Many Miglets are at home with their parents after getting their degrees, because our corporations have discovered that it is much cheaper to hire Miglets in India or China instead. Meanwhile, the student loans of the Miglets keep piling up. Besides, the salaries in India are also being driven down by the paucity of jobs there as the spiraling race to the bottom is well and truly underway for the Miglets.

There are many intellectuals around the world who believe that this situation is inevitable because the optimum human population of planet Earth is 1-2 billion[14]. But it seems to me that they base their estimates on the current Western way of life or on the life that they are personally leading.

Suppose that our industrial civilization had arisen out of China instead of the West. And imagine that a Chinese "McDonalds" popularized the consumption of Bear-Paw soup and Shark-fin soup throughout the world, instead of the Hamburgers and Fillet-of-Fish that we find peddled in every corner of the globe today. Bear Paw soup is a Chinese delicacy with strong parallels to shark-fin soup. To make bear-paw soup, we cut off the paws of a bear and exhibit the bear outside our restaurant with its bleeding hands so that our customers know that the soup inside is fresh. Only the paws of the bear are used to make the soup while the rest of the bear is discarded after it bleeds to death.

To make shark-fin soup, we cut off the fins of the shark and let the rest of the shark die in the water.

Then the common diet around the world would have involved an order of magnitude more waste than the diet of Hamburgers and Fillet-of-fish that we consider to be normal today. In that case, perhaps, at most 100 million people could be supported in such a lifestyle on this planet. We would have to grow crops to feed small animals to feed the bears, just to cut off their paws and use the paws for human food. We would have to grow crops to feed small fish to feed the sharks, just to cut off their fins and use them for human food.

In such a world, even by 500 BC at the time of the Buddha, the human population would have been too many and completely non sustainable! Therefore, I believe that the intellectuals who contend that the optimum human population on the planet is 1-2 billion are not asking the right question. The question shouldn't be, "What is the optimum human population on the planet if everyone lived like us?" Instead, it is much more productive to address the question, "How should we be living, given that there are 7 billion people and 20-100 million other species on the planet today?"

Just as there is a major disparity in overall consumption between the top and the bottom, there is a similar glaring disparity in fossil energy consumption as well. Fossil fuels such as coal, oil and natural gas are all carbon-based fuels that release energy when they are burnt, but they also convert the stored carbon on land into the gas, Carbon DiOxide (CO_2), which accumulates in the atmosphere, blanketing the earth and causing it to warm. Prof. Stephen Pacala of Princeton has calculated that the top 500 million people on the planet are responsible for half the carbon emissions and that these people are actually everywhere[15]. They aren't just in the global North, but they are also found among the top consumers in India, China, Brazil, Russia and the global South. But when we consider consumption on a country by country basis, these top consumers in the global South are able to hide behind the skirts and saris of the poor and present themselves as holier than thou to the rich countries. As a result, negotiations at the United Nations (UN) on controlling atmospheric emissions from burning fossil fuels have been far from effective, because this is truly a grassroots behavioral problem that is not conducive to a country-by-country top-down solution. In our globalized economy, the top consumers in the global South are just as responsible for the world's environmental problems as the consumers in the global North.

At the UN Earth Summit in Rio de Janeiro in 1992, President George H.W. Bush of the United States flatly asserted to the world[16], "The American Way of Life is Non-Negotiable." This didn't win America many friends in the world community. Said Jayanthi Natarajan, the new Minister for the Environment of India[17],

> "There is something fundamentally unfair about countries that have used up all the natural resources and reserves on our planet, turning around and preaching to us about reducing our carbon footprint when with our billion-plus population, we are not even a blip on the radar of carbon emission. Western countries who preach the most have absolutely no intention of cutting their own carbon emission or to rethink about their wasteful economies..."

Meanwhile, the present American way of Life is being exported by corporations to people all over the world. For corporations are not particularly sensitive to where their profits are derived from and want to keep growing them regardless of the consequences to the planet. They may be American corporations, but their loyalty to their bottom line invariably exceeds that to America or to the planet.

And so it goes. The rich people in the world want the poor people to stop having kids, while the poor people want the rich to stop consuming so much. Both sides are mostly interested in getting other people to change.

But the only way to get other people to change is to first change ourselves, to undergo our own personal Metamorphosis.

1.2 Civilizational Transformations

It is always the Caterpillar that has eaten its fill that is ready for the final Metamorphosis into a Butterfly. As a species, if we recognize that we need to become Butterflies, then clearly the onus is on the well-off people to take the lead. Specifically since you are reading this book, you are definitely a candidate to undergo Metamorphosis regardless of where you live, if you aren't doing so already. Such a mass civilizational transformation accompanied by a change in human consciousness is surely necessary to turn things around.

As Jeremy Rifkin has pointed out, in the past, mass civilizational transformations such as this have always been accompanied by a revolution in the energy infrastructure used to fuel the civilization along with a revolution in the means of communications[18]. The original civilizational transformation occurred when humans harnessed fire and began to cook food. This increased the variety of foods available to human beings, for example, by making meat edible, and greatly fueled a growth in human population. Therefore, fire on-demand through the burning of wood was the original energy revolution. The corresponding communications revolution was the invention of language.

Agrarian societies developed when humans were able to harness the energy of animals to plough their fields to plant crops. The agricultural revolution was fueled by Animal energy and the corresponding communications revolution was the invention of writing. Just as cuneiform emerged in Sumeria, hieroglyphics in Egypt, writing emerged in the Indus Valley, in the Yangtze valley in China and in Mexico independently so that agrarian societies could tabulate their crop yields and transact with each other. This fueled the growth of cities where people were freed from the burden of procuring food for their sustenance and started exploring other intellectual pursuits such as music, art, poetry and philosophy.

The First Industrial revolution was fueled through the burning of coal and the invention of the steam engine to harness the energy of coal. The communications revolution that accompanied this was the printing press that allowed for the rapid dissemination of information among the learned classes, which led to a great acceleration of scientific knowledge and a rapid growth in human population as death rates began to decline. One of the greatest contributions of Western civilization to humanity, the invention of the Scientific Method, occurred in this era.

The energy infrastructure transformation that triggered the Second Industrial revolution was the use of oil, natural gas and uranium. The corresponding communications revolution occurred through the invention of telephony, telegraphy, radio and television.

Now we are in the midst of a great revolution in communications with the advent of the personal computer, the internet and wireless communications technologies. What Jeremy Rifkin calls the distributed communications revolution is already in place as individuals blog, post videos and information online and act as distributed news sources. With social networking, individuals are also reaching well beyond their immediate circles and can potentially influence opinions worldwide. While all previous communications revolutions increasingly aggregated power in the hands of a few, this time it is fundamentally different as individuals are empowered. While it is true that the distributed communications infrastructure is still being controlled by a few corporations or the power elite and their intent in providing this infrastructure is to reach out and persuade us to buy things that we don't need, nevertheless, the infrastructure is there for us to utilize. As Michael Moore, the documentary filmmaker said[19], "Everybody is a filmmaker now. Everybody has a camera!" And the poor people on the planet, who had not even enjoyed the benefits of the printing press, the communications revolution of the middle ages, are suddenly leap frogging everything and using cell phones and wireless technology.

As Kishore Mahbubhani wrote in the Financial Times[20],

"Dictators are falling. Democracies are failing. A curious coincidence? Or is it, perhaps, a sign that something fundamental has changed in the grain of human history. I believe so. How do dictators survive? They tell lies. Muammar Gaddafi was one of the biggest liars of all time. He claimed that his people loved him. He also controlled the flow of information to his people to prevent any alternative narrative taking hold. Then the simple cellphone enabled people to connect. The truth spread widely to drown out all the lies that the colonel broadcast over the airwaves. So why are democracies failing at the same time? The simple answer: democracies have also been telling lies."

A distributed energy revolution is also underway with solar and wind energy solutions getting deployed world wide, but it hasn't attained the same reach as the distributed communications revolution yet. When people are able to tap into the energy that's literally falling on their heads and blowing on their faces everywhere, they will truly become untethered from the clutches of another and be free to act independently. Once again, the poorest people on the planet have the opportunity to leap frog all the past energy revolutions that never reached them, coal, oil, natural gas and uranium, and move straight from animal power to the solar age.

Our present world situation is ripe for the Metamorphosis, the civilizational transformation that can occur on the basis of the distributed energy and distributed communications revolutions. As individuals are empowered and hierarchical power structures are weakened, the Butterfly can emerge en masse, as our expectation of what constitutes an average human being in society changes. Until now, all major civilizational transformations had increasingly separated humans from Nature, because they had increasingly concentrated power in a few elite institutions or corporations. But the Metamorphosis aims to reconnect humans back to Nature, so that the Butterfly can help repair and heal the damage that was done in the past. To facilitate this transformation, a new political mindset is emerging among the Miglets who grew up on the internet and socialized media. They are no longer interested in the authoritarian, Right vs. Left type judgmental, hierarchical political structures of the past and are increasingly seeking the distributed, collaborative political structures of the future.

1.3 Who are We?[21]

Reconnecting humans back to Nature is about reconnecting humans back to reality by identifying and overturning all the absurd notions that underly our current civilization. It is also about reconnecting humans back to our own selves. Nature isn't just out there in the wilderness, Nature is within us as well. Put it another way, to dissolve the illusion that we are separate from one another and from Nature, it is necessary to dissolve our ego which promotes our illusion of separateness and thus connect within to the Atman, the Spirit within us. To become compassionate towards all Life and thus fulfill our purpose as human Butterflies, we must first practice to be self-compassionate. As such, the Metamorphosis is a deeply spiritual undertaking.

The foundational texts of Hinduism are the Vedas. In the Vedic view, the universe is built up of dualities. And human beings are no exceptions. The Bhagavad Gita, the Hindu song of songs, speaks explicitly of the ego/Atman or ego/Spirit duality. Perhaps a more comprehensible model is the Caterpillar/Butterfly duality described above, but the gist is the same.

It is the identification with the ego that drives the Caterpillar side of a human being. The Caterpillar is driven by the fear that arises out of a sense of separation and isolation.

It is the identification with the Atman that drives the Butterfly side of a human being. The Butterfly is driven by the love that arises out of a sense of belonging.

According to the Upanishads, which are a commentary on the Vedas, we are each built with the exact same ingredients, with the exact same potential for hate and for love, for crassness and for greatness. It is up to us to choose who we want to be.

Moments in time when we feel alienated, separated, angry and hurt strengthen our ego. The stories that etch these moments in our memory then begin to further cloud our perception of reality. That is, until we become aware of these filters and train ourselves to see through them.

Moments in time when we feel perfectly at peace with the universe and experience a sheer sense of joy at being alive, strengthen our identification with the Atman. It used to happen to me occasionally; the first time I saw the Himalayas from the foothills of Nainital in the northern state of Uttar Pradesh in India, the first time I saw the Grand Canyon in Arizona, the joyful moments I spent with family and friends, the moments I held my children in my arms and the moments when I found a creative solution to some knotty problem or the other. Fortunately, it now happens to me a lot more frequently, every time our granddaughter, Kimaya, tries to gnaw on my nose with her baby gums, and every time when I meditate. These are the moments when our sense of separation dissolves and we feel at one with the universe.

Just as the ego is experienced as an accumulation of separation events, the Atman is experienced as an accumulation of such integration events. In our current scientific world view, we are biased to recognize just the ego as the totality of our human experience and to ignore the Atman. We assign superiority and grade intelligence on the basis of the extent to which beings can separate themselves and stand out from the whole, to the monkey that can recognize itself in the mirror, but not to the spider that can delicately spin an intricate web that captures the right sized insect for its specialized diet. We studiously ignore the intelligence in Nature that arises in the integration, ascribing that intelligence to a machine-like, automatic "instinct." And we celebrate the intelligence that separates a being from its environment. We celebrate the ego. The problem was compounded by Sigmund Freud, one of the more influential thinkers of the 20th century, who proposed that humans are born with just the Id in the mind, a self-centered construct that is only interested in the fulfillment of its desires that then develops into the adult ego. He based this on his observation that a baby only significantly interacts with the universe when it cries. And it cries only when it desires something, a change of diaper, some food or some sleep.

So in the Freudian world view, babies are born Caterpillars.

Sigmund Freud had perhaps never played with his baby daughter in the vast intervals between her crying when she could have gnawed on his nose with her baby gums and smiled up at him. Perhaps she only cried when she needed something, some food, a change of diaper, some sleep and not because she desired anything. But it was Freud's ideas that were developed into the world consumer societies of the 20th century by his nephew, Edward Bernays, with the encouragement and support of large corporations. Bernays originated the idea of linking products to emotional desires and feelings in order to persuade people to behave irrationally, to consume things that they really didn't need. He persuaded women to take up smoking even though it was taboo until then, by convincing them that the cigarette was a phallic symbol of power. He dreamed up many of the techniques of mass consumer persuasion using celebrities that we now experience on a daily basis. The incessant fulfillment of desires, supposedly in the pursuit of happiness, was born in what the British Broadcasting Corporation (BBC) has characterized as the "Century of the Self," the century of the ego, the 20th century[22].

In November of 2009, I was at an India Development Coalition of America[23] (IDCA) conference in Chicago to talk about our non-profit work in India. At the end of the conference, my host, the President of IDCA, Dr. Mohan Jain, invited me to visit the Hindu Swaminarayan temple in Bartlett, IL, with him to meet with a holy man, Swami Atma Swarupji, from New Delhi, India. He said that if the Swamiji got interested in our work, then he could bring an enormous amount of resources to fund our project. That temple in Bartlett sported a huge dome plated in gold! Since my flight back to San Francisco was the next morning, I readily agreed and went with him to the temple. Once there, Dr. Jain approached the temple priests, dropped some names and got us into the anteroom of the Swamiji's chamber, where we were told to wait until the Swamiji finished his meeting inside. Waiting with us was a wizened man, Mr. K. R. Jani, probably in his eighties and whose son was in the meeting with the Swamiji. We were chatting with him for a while about this and that until we were told that the Swamiji was running late for the assembly and that we should have our audience with the Swamiji the same way as everybody else, after his discourse to the assembly in the cavernous main hall. Since we were honored guests, we would be given front row seats during the discourse and would get to see the Swamiji at the head of the line.

As we were returning to the main hall, Mr. Jani walked alongside me holding my hand. He became very solicitous of my work and offered me words of encouragement. Then he took me past the entrance to the hall, gave me a great bear hug and held my head in his hands. He had one thumb in the center of my forehead and another at the back, while he said, "This is Sri Krishna (God) speaking to you through me. Your inspiration comes through here (pressing the back of my head) and your execution comes through here (pressing the front of my head). Do your work without any ego and let me do it through you." Then he turned away and left.

Of course, what Mr. Jani said was straight from the Bhagavad Gita where Sri Krishna tells his devotee, Arjuna, to perform fruit-forsaking action without any ego and thus become the flute that Sri Krishna can play to create melodious music. This is the very symbolism of the statues of Lord Krishna playing the flute that we commonly find in Indian bazaars. That didn't surprise me. What surprised me was that I was left in a daze with his thumb imprinted on my forehead for the next few hours and I still feel his thumb on a daily basis when I meditate. Egoless, fruit-forsaking action is what the Gita calls for, but the Century of the Self celebrated the exact opposite, the ego-driven, fruit-seeking action.

The Gita proposes that only the Atman within us is permanent, while all other aspects of the human experience including the ego, is impermanent. The Atman can be viewed as a ripple in an eternal ocean of Consciousness that takes up temporary residence in a material body that is impermanent and constantly flowing. The Gita states that the Atman never dies and returns back to the ocean of

Consciousness when the body dies. The body is controlled by a mind and it gets its inputs and performs its actions through its various senses. The Gita distinguishes the mind (the "subconscious mind" in Psychological terms) from the discriminating Intellect (the "rational mind" in Psychological terms), which is said to reside behind the center of the forehead, in the pre-frontal cortex of the brain. The ego then completes the makeup of the human being, leading to a sense of separateness from the universe. The ego is fluid and impermanent as it can be diminished through practice and awareness.

There is a scientific basis for the Gita's model of the Atman being a ripple in the eternal ocean of Consciousness. The microbiologist, Lynn Margulis, has accumulated compelling evidence that the mitochondria in human and other animal cells and the chloroplasts in plant cells are all bacterial in origin and therefore, each complex being can be viewed as a closely cooperating colony of trillions of bacteria[24]. Therefore, we are each, literally and physically, part of the whole and deeply embedded in it, except that we seem to have lost our connection with the whole over the years due to our cultural upbringing.

According to the Gita model, the Caterpillar is enslaved by the ego. The Caterpillar has a fluid mind that is constantly invoking new desires and commanding the intellect to find ways to satisfy those desires. In a Caterpillar, the intellect is the servant of the mind, which is the servant of the ego.

Thus, the Caterpillar is oriented outwards in service of the ego. Its purpose is to fulfill its never-ending desires.

In a Butterfly, the ego is diminished and even non-existent while the mind is steady and solid, well controlled by the discriminating intellect. In a Butterfly, the mind is a servant of the intellect.

The Butterfly is oriented inwards in service of the Atman and through it, all Creation. Thus, it has a purpose larger than itself, which becomes the source of its deep-rooted happiness.

Caterpillars are driven by fear, specifically their fear of Death.

Butterflies are driven by love, specifically their love of Life.

The Caterpillar/Butterfly dichotomy is thus intimately tied to the fear/love duality which resides in the Medulla oblongata or the reptilian brain of most complex beings. This is fundamental to most life forms as we all need fear to escape a predator and we all need love to take care of our babies. Without these two fundamental functions, a species would find it hard to survive. In her popular Ted talk[25], Dr. Jill Bolte Taylor, the Harvard-trained neuroanatomist, describes the left-brain/right-brain dichotomy that arises out of this fear/love duality and that she vividly experienced as a result of a stroke which temporarily disabled the left hemisphere of her brain. The right hemisphere of her brain, which is devoted to processing present moment reality, made her feel intensely connected with the universe around her. The left hemisphere is about judging the present moment using stored memories from the past, to filter out the enormous inputs received by the brain at each instant. It is also about making predictions and plans for the future. The ego resides in the past and in the future and dissolves in the light of present moment reality. Thus, the Caterpillar is ruled by the left hemisphere of the brain while the Butterfly empowers the right, while using the left as a tool.

And it is my thesis that in our modern industrial culture, we deliberately train all our babies to grow up to be Caterpillars, even though they are born Butterflies.

2. Karma

"Nobody made a greater mistake than he who did nothing because he could only do a little" - Edmund Burke.

Imagine the world that we want to live in. Then act to make that world a reality.

The trouble with human beings is that most of us act without thinking about the world that we are creating through our actions. Very few of us would imagine or desire a world without tigers, lions, elephants, birds, forests and fishes. Very few of us would consciously act to make such a lifeless and desolate world happen. That in fact, we are acting to make such a lifeless, desolate world happen is therefore, an abject failure of our imagination. Either that, or we are disconnected from our real desires and needs and are acting in a state of stupor.

But actions have consequences.

And inaction has consequences.

That in essence, is what the ancient doctrine of Karma is all about.

Karma is not about judgement. It is about simple physics and chemistry in a chaotic, deeply interconnected system that is our Earth. In such a system, seemingly small actions can have large consequences. So can inaction. A tiny butterfly flutters its wings in the Amazon and stops a hurricane from forming over the Caribbean. A woman's stubborn refusal to move to the back of the bus causes a social upheaval in the segregated, deep South in America.

It is impossible to predict which action or which inaction causes the chaotic system to tip over to a new state. Therefore, in the face of climate change and the environmental catastrophes looming over Life on Earth, we must encourage a million Rosa Parks' to refuse to budge and a billion butterflies to flutter their wings. That is the only way for the system that is driving modern industrial civilization over a cliff to reverse course and right itself.

2.1 Actions and Inaction Matter

The promise of Karma is that no act is ever ignored by the universe. Every act matters, no matter how insignificant we may regard them to be as we perform them. The repercussions from our actions ripple through the fabric of the universe, perhaps well beyond our own individual lifespans. Therein lies the "Karmic retribution" of everyday lore.

The Bhagavad Gita is very clear on the point that we are given the gift of Life to act, for Life is about action. I have so much to be grateful for in this gift of Life. I am now in the 9th year of my bonus years on this Earth. Born in 1960 in India, my life expectancy at birth was 42 years. My twin perished at birth, perhaps to allow me this extra statistical lease on Life.

When India achieved independence from the United Kingdom (UK) in 1947, the life expectancy at birth for an Indian infant was a mere 36 years[1]. The corresponding figure for an infant born in the UK was 65 years. Childhood vaccinations were almost non-existent in India and childhood malnourishment rates were approximately 90%, leading to high infant mortality rates, which was the main reason for the low life expectancy in India compared to the UK in 1947. As a result, the average Indian woman bore 5.6 children since many of her children did not live past their first year. When I was a four year old, I distinctly recall my grandmother advising my mother not to get too attached to my youngest brother until he was at least a year old.

For my grandmother grew up in the colonial era, expecting babies to die within their first year. But my siblings and I received our childhood vaccines plus adequate nutrition and survived.

In the UK of 1947, the average woman bore only 2.7 children.

Fast forward to 2009 and the life-expectancy of a child born in India is 64 years, the childhood malnourishment rate has fallen to around 48%, while the fertility rate has declined to 2.68 children per woman. Thus, sixty-two years after being freed from the yoke of colonialism, India is on a demographic par with the UK of 1947, but with a population that has ballooned to almost 1.2 billion.

Not only do actions matter, but inaction matters as well. The colonizers' neglect of infant mortality, childhood diseases, childhood malnourishment and the general standard of living among the colonized now matter as there are billions more mouths to feed on the planet than otherwise. Our continued neglect of public health in poor communities matter, not just for the poor themselves, but for the rich and for Life on Earth, in general.

The five decades of my existence since 1960 have been the most profoundly transformational years in the anthropogenic Earth's history. Half of the destruction on the planet has occurred in these years, equalling the destruction wrought by the previous 500 generations of humans combined[2].

As a witness to this destruction, I can safely assert that our world is in a mess.

Except for the rarest among us, our lives are in a mess as well.

It is my thesis that the two are inextricably linked, that in order to fix our world's mess, we must help each other fix the messes in our respective lives, while taking off the cultural blinders that drive us to be so destructive, to behave like Caterpillars. This is not a new insight as mystics from time immemorial have been steadfast in the belief that though everything is in a mess, all is well. The world doesn't need to be changed, but it just needs to be understood. The world is behaving precisely as it should, generating consequences for all our actions and inaction without prejudice or judgement, and it is perfect. It is we who need to change. And, when our understanding improves, we will automatically change, effortlessly, and the world will react accordingly.

Though a young species that has barely been around for two hundred thousand years, and engaged in industrial activity for just over two hundred of those, human beings have been furiously transforming the planet on a vast scale. The statistics are stark. More than two-thirds of the ice-free land area of the planet has been converted into desert, pasture land, urbanized land or agricultural land. Half of the world's forests have been destroyed. About 30 million acres of tropical forests are still being razed down each and every year, at a rate of 1 acre every second. On the flip side, around 10 million acres of agricultural land is being abandoned each year in temperate regions, which Nature does reclaim as forest land, but newly regenerated forest cannot begin to compensate for the diversity lost in the old growth forests that are being cut down by humans. In the ocean, around 90% of predatory fish stocks have disappeared. They were mostly eaten just in the past fifty years. More than three quarters of all marine fisheries have been overfished. In the history of Life, species have evolved and have become extinct at a fairly slow pace. There have been instances in the fossil record when species have gone extinct at a much accelerated pace and those have resulted in mass extinction events. Our present era rivals such instances. Overall, species are now going extinct at 100-1000 times the long term background rate. And the rate of extinction has been increasing by a factor of 10 roughly every 20 years. According to the American Museum of Natural History, we are now in the midst of the fastest mass extinction in the Earth's 4.5 billion year history and unlike prior extinctions, this Sixth Great Mass Extinction event happening today is mainly the result of human actions and not of natural phenomena[3].

Our actions matter.

It does not take complex arithmetic to understand where all this will lead to and how much time we have to make the necessary changes. If half of the world's forests were destroyed by humans mainly within the past fifty years, how many years do we need to destroy the rest if the rate of deforestation continues to grow at the same pace? If 90% of the predatory fish stocks were eaten mainly in the past fifty years, how many years do we need to eat the rest if our rate of consumption continues to grow at the same pace? If 97% of the tigers were killed off in the wild in the past 100 years, how many years do we need to kill off the remaining 3%?

So much for the carbon based life forms that presently inhabit the earth. In addition, billions of tons of fossilized life forms are being disinterred from the earth and burnt up annually, spewing toxic chemicals and greenhouse gases into the environment, turning it into a toxic soup. That toxic soup along with the radiation leaks from nuclear disasters such as Fukushima Daiichi in Japan, is now making its way up the food chain, ruining the health of the planet's complex biota, including humans.

None of these facts are in dispute. Nor is it in dispute that humans have been doing all this destruction and creating all this pollution, and mostly over the past fifty years. Life is dying before our very eyes at such a pace that one of the leading biologists in the world, Prof. Edward O. Wilson of Harvard University, felt compelled to write an entire book, "The Creation[4]," with the subtitle, "An Appeal to Save Life on Earth."

I found this absolutely shocking! Here's this gentle, octogenarian human being appealing to his fellow human beings to save Life on Earth!

Now who among us, in their right minds, truly wants to destroy Life on Earth? That's about as dumb as sawing off the tree limb that we're sitting on. Yet we seem to be doing it collectively, as if that was the intended purpose of our modern industrial civilization, this global consumer culture that we are ensconced in.

This book is not about assigning blame to this generation or the previous or to this country or that for creating this mess. Billy Joel said it best when he sang: "We didn't start the fire. It was always burning since the world's been turning." However, this book is about showing that we are certainly stoking the fire until it is threatening to engulf the whole planet and consume us all. That it is the responsibility of those alive today to start reversing the process, to halt the Holocaust on Life that we've triggered through our actions. And that it is no longer enough to excuse ourselves by singing that the fire has always been burning, but to pick up buckets and form brigades to bring it under control, beginning with the fire engulfing ourselves.

2.2 The Strange Paradox

Here's what's amazing: the greatest story to ever unfold on the planet, our imminent march over a cliff following an invisible Pied Piper, is playing out in slow motion while the mainstream media seems to be strangely apathetic, especially in the United States. As if it has also been drugged into a state of stupor.

The strange paradox is that even as we are marching in lock-step following the Pied Piper, we're living better and better in most places on Earth, at least among the affluent community. In general, the material well being at the top has never been this good. Weddings in Mumbai, India, are being celebrated with lavish buffets spanning two city blocks with cuisine from all over the world served in separate stations. But a billion people are also going hungry over the same world, including a few million people within a few miles from these weddings. Just as Charles Dickens famously wrote[5], "It was the best of times, it was the worst of times," referring to the late 18th century, it is tempting to think that it has always been this way and therefore, there is not much that we need to do to correct the situation. That it will continue just the same for ever and ever. After all, even as problems arose in the past, we have always overcome them through the sheer power of our technological ideas and thereby defied past Malthusian predictions of a civilizational collapse.

The Reverend Thomas Robert Malthus[6], an English philosopher who lived in the late 18th and early 19th century, calculated that as population increases exponentially while food production increases arithmetically, a civilizational collapse was inevitable in the future. But as a Malthusian deadline loomed in the early 20th century, the German scientists, Fritz Haber and Carl Bosch devised a process[7] for converting atmospheric nitrogen into fertilizer, thus ushering in a rapid improvement in the productivity of agricultural land, thereby buying humanity an additional few decades of business as usual existence. Later, when another Malthusian deadline approached in the 1970s and 1980s, Norman Borlaug unleashed his Green Revolution[8], the improvement of agricultural productivity through improved seeds, irrigation and modern production methods, and averted that imminent disaster.

Perhaps technology is the drug that has numbed the media pundits. They are probably expecting another just-in-time technological miracle that will avert catastrophe and which they can report upon breathlessly. However those past technological solutions have all turned out to be narrow-sighted fixes in retrospect, each with its own adverse, unintended consequences. The continued destruction of the environment is testament to the temporary nature of those past fixes. The Haber-Bosch process has resulted in a tremendous imbalance in the planet's nitrogen cycle, leading to vast oceanic dead zones at the mouths of rivers[9]. And the Green Revolution has depleted underground aquifers and the vitality of soils world wide[10]. In other words, we grew more food for humans by depleting the capital wealth of the planet and by killing other species over the past century, by favoring a few select and even genetically modified species over the biodiversity of the Earth. And Nature is now grinding out her consequences.

In his book, "The Bridge at the Edge of the World[11]," Gus Speth, the Dean of Environmental Studies at Yale University, puts it succinctly,

> "All we have to do to destroy the planet's climate and biota and leave a ruined world to our children and grandchildren is to keep doing exactly what we are doing today, with no growth in human population or the world economy. Just continue to generate greenhouse gases at current rates; just continue to impoverish ecosystems and release toxic chemicals at current rates; and the world in the latter part of this century won't be fit to live in. But human activities are not holding at current levels - they are accelerating dramatically. The size of the world economy has more than quadrupled since 1960 and is projected

to quadruple again by mid-century. It took all of human history to grow the $7-trillion world economy of 1950. We now grow that amount in a decade."

In his recent book, "The Great Disruption[12]," Paul Gilding flatly asserts that this projected four fold growth in the economy by mid-century, isn't going to happen. He makes a compelling case that a Great Disruption of modern industrial civilization already began with the financial collapse of 2008 and that within the next decade or so, humans are going to be engaged in an all out, World War II style, effort to merely survive as the planet reacts furiously to our past depredations.

Only time will tell. But it is important to note that in the past, those civilizational crises were averted not because the Malthusian projections were wrong, but because people cared enough to take action.

2.3 The Axiomatic Flaws

When things are going so spectacularly wrong in any system, it is a safe bet that there is a fundamental flaw in one or more of the foundational axioms of that system. If we identify and correct these flaws, then the system has a chance to right itself. But if we keep patching up the symptoms that arise in the faulty system, then we will be applying more and more complex patches over time until the whole system collapses spectacularly as well. It is not a question of whether such a collapse occurs, but when. For the modern human enterprise has truly been a seat-of-the-pants engineered, waste producing, rickety contraption when compared to the robust, waste-free ecosystems that Nature has developed through eons of evolution.

Thus far, we have been applying patches to the symptoms of our systemic distress as they arose. Both the Haber-Bosch process and Norman Borlaug's Green Revolution were patches. It is long past due for us to examine the foundational flaws in the human enterprise instead, even if this results in a fundamental sea change in the way we live out our lives. And it is up to us who are alive today to make that shift. And it is most likely the Miglets that will lead this social transformation, the Metamorphosis.

In contrast, members of my generation that is currently in power throughout the planet are still behaving cynically, perhaps because we have more years behind us than ahead of us. We have oinked our way through the true wealth of the planet, the totality and diversity of Life on the only life-supporting planet in our galactic neighborhood, leaving a vastly diminished base for the Miglets and future generations to work with. And though in power, we have yet to face up to it. We have punted the problems down the road at every opportunity, while looking for make believe fixes of the technological kind. We have found trillions of dollars to bail out the banks and the bonuses of their executives, but we have failed to find the resources needed to ensure a livable planet for our children. We have been behaving like addicts, desperately seeking the next fix to keep our destructive enterprise going. We promised change to energize the Miglets during the 2008 American presidential elections and then when ensconced in power, we delivered more of the same.

How did it come to this?

The mythology that drives our global, modern culture stems from the rousing words of Thomas Jefferson and the founding fathers of the American Revolution: "We hold these truths to be self-evident, that all men are created equal, that they are endowed by their Creator with certain unalienable Rights, that among these are Life, Liberty and the Pursuit of Happiness[13]." The "Unanimous Declaration of the thirteen United States of America" is probably one of the most consequential documents ever produced by human beings, but it has been profoundly misinterpreted to the point where we are currently doing the opposite of what it says. I base this observation upon my familiarity with the Bhagavad Gita, but I'd wager that a scholar in any other religion would be able to point to the same parallels in their foundational texts as well.

As the science fiction author, Philip K. Dick, wrote, "The basic tool for the manipulation of reality is the manipulation of words. If you can control the meaning of the words, you can control the people who must use the words[14]." And this manipulation of words began even as the Declaration of Independence was being written. With respect to the phrase, "all men are created equal," the English abolitionist, Thomas Day, wrote in a 1776 letter, "If there be an object truly ridiculous in nature, it is an American patriot, signing resolutions of independency with the one hand, and with the other brandishing a whip over his affrighted slaves[15]." Indeed, many of the signatories of the Declaration of Independence were slave owners. Later, in the Lincoln-Douglas debates of 1858, Stephen Douglas asserted that the Founding Fathers only intended the phrase "all men are created equal" to be applied to white men only, that the purpose of the Declaration was to justify the independence of the United States and not to proclaim the equality of any "inferior or degraded race[16]." But Lincoln took the expansive view that the language of the Declaration was deliberately universal, setting a high moral standard for which the American republic should aspire. When Lincoln's view won the argument, the arc of the moral universe was bent towards justice as Rev. Martin Luther King, Jr., so eloquently put it[17].

But the much more egregious misinterpretations of the Declaration of Independence that bedevil our modern industrial civilization lie in the "rights" to Life, Liberty and the Pursuit of Happiness. In America, the right to Life is variously interpreted to argue against abortion rights in some quarters and against capital punishment in other quarters. But in our modern culture, the ideal Life itself has been interpreted to mean the attainment of leisure. In our popular media, we celebrate those who have achieved a life of leisure and indolence.

The Bhagavad Gita stridently opposes such a Life of indolence and inaction. "Do not be attached to inaction[18]," warns Lord Krishna to Arjuna. There is no doubt about it; the Gita is the gospel of action. We are given the gift of Life in order to act. And our actions matter profoundly.

Next is our interpretation of "Liberty." According to the Bhagavad Gita, true liberty is attained when we are steady-minded enough so that neither praise nor criticism affects us and what we think, say and do are all in perfect harmony. Then we are truly untethered from the control of others and are free to pursue our autonomous path, to dance our dance. Unfortunately, in our current culture, praise and criticism are used to train humans from their infancy onward to adhere to societal norms, while a steady programming of fear and violence is used to keep them from their true expressions, to turn them into compliant Caterpillars. The right to Liberty has been misinterpreted to mean the right to bombard an individual with manipulative messages through the mass media, while individual liberty is mistaken to be our freedom to do any imaginable depredation on Nature. The right to Liberty is now the right to destroy the property that has been designated to us by the society we live in, while being thoroughly controlled by the punishment and reward systems of that same society.

Finally, the biggest mistake of all has been our misinterpretation of the Pursuit of Happiness. In our modern culture, this has been completely conflated with the Pursuit of Affluence in order to fulfill an infinite procession of desires that are being constantly stoked through the media. The Gita clearly states that true happiness is found within ourselves, and that the incessant fulfillment of desires leads to the exact opposite of happiness. This final misinterpretation is what gets even well-meaning politicians to pay instant attention to the misfortunes of corporations and their executives in their drive to keep the affluence engine chugging along, while ignoring the needs of the Miglets and future generations.

Thus, we have become a culture that has profoundly misinterpreted its foundational axioms and is presently doing the exact opposite of what is in the ancient texts. Perhaps, if we interpret our foundational axioms following the guidance in these texts and act accordingly, our world and our lives wouldn't be in such a complete and utter mess.

Just like the character, George Costanza, in the popular TV serial, "Seinfeld," we would be better off doing the exact opposite of everything we do today.

Carbon Dharma

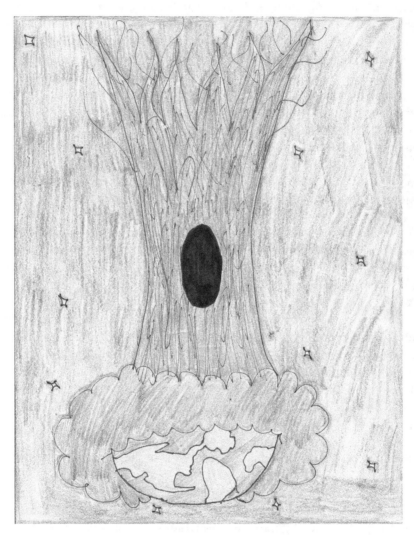

Fig. 2. *The Cosmic Fig Tree, a metaphor for the
unsentimental wish-granting universe.*

3. Dharma

"Your duty is to act, but not to reap the fruits of your actions. Do not be attached to the rewards for your actions, and do not be attached to inaction" - Bhagavad Gita 2:47.

If all humans ceased to exist right now, Earth and its other inhabitants will have a good chance to recover and Life will most likely flourish once again. Alan Weisman wrote an entire book, "The World Without Us[1]," based on just such a premise. Therefore, it is our daily human actions that are contributing to the utter mess that we find ourselves in, drip by drip. Perform the right actions and not only can we live in harmony with Nature, but we can even help Nature recover from our past depredations. Perhaps then, Life can flourish once again on Earth, but with us as a part of it. If the insects, the birds, the fishes and the animals know how to take the right actions and live in harmony with Nature, aren't we humans smart enough to organize our systems and actions to do the same?

And that's where the ancient concept of "Dharma" comes in.

While the doctrine of Karma deals with actions and the consequences of actions, the concept of Dharma deals with the choices facing humans at every moment in life regarding what actions to take. It addresses the thorny question of how to take the "right" action in any given circumstance. How do we determine what is the right thing to do? Dharma can be loosely translated as "righteousness," but it does not carry the negative connotations of that English word. The Hindu epic, Mahabharata, deals with numerous gray areas that individuals encounter during the course of their lives and uses parables and tales to illuminate the path of Dharma under those circumstances. It implicitly acknowledges that there may not be universal recipes for choosing the right actions under every circumstance. Composed entirely in Sanskrit verse to aid in the oral tradition for transmitting the epic from generation to generation, the Mahabharata is seven times longer than the Iliad and the Odyssey put together, though less well known in the Western, industrial world.

The Mahabharata is centered on the Battle of Kurukshetra that is set in ancient India of 5100 years ago. The Bhagavad Gita is the very essence of the Mahabharata and it is better known in the Western world. The Gita is composed as a question and answer session between Lord Krishna and Arjuna regarding Dharma. As the story goes, Arjuna, the preeminent warrior in the Mahabharata, is torn between his duty to fight on behalf of his brothers to restore the kingdom of the Pandavas versus his distaste for killing his cousins and elders in the opposing army of the Kauravas and therefore, seeks counsel from his charioteer, Lord Krishna. Arjuna is experiencing what is known as "Dharma Sankata," the dilemma that occurs when there are no clear choices available before us. Though the Bhagavad Gita is written as a discourse that eventually persuades Arjuna to plunge into the battle that follows in the story, it was also the source of inspiration for Mahatma Gandhi to wage his non-violent, civil disobedience movement that eventually dislodged the British from India. For the principles of Dharma espoused in the Gita and in the Mahabharata cut both ways.

3.1 The Cosmic Fig Tree

Perhaps the quintessential story to illustrate the framework for Dharma is the Cosmic Fig Tree (Kalpataaru) story from the first chapter of the Rig Veda, the first of the foundational texts of Hinduism[2]. Also related by Lord Krishna in the Bhagavad Gita, this story begins with some children playing with their sticks, stones and rag dolls on the floor of their hut when the proverbial rich uncle visits them. The uncle tells the children, "What are you doing playing with these things when the Cosmic Fig Tree is right outside your hut? Go outside and wish for whatever you want under the tree and it will give it to you. Then you can be playing with real toys instead of these trifles."

The children don't believe him. How could it be possible that such a wish-fulfilling tree even exists? So they wait until the uncle leaves and then they rush out to the tree and start wishing.

They wish for sweets and lo and behold, they get them! But they gorge on the sweets and get stomach aches. They wish for fancy toys and they get them. But they play with those fancy toys and get bored. Fancier toys lead to greater boredom. What they didn't realize was that the tree always granted wishes in dualities: what was wished for, along with its built-in opposite. They had to accept both at the same time, for the universe was built up of such dualities only. The net result was that the children were miserable throughout their childhood, but they couldn't stop wishing under the tree.

Then they became adults and they still came to the tree and wished under it all the time. Now it was Sex, Fame, Money and Power, the four main fruits of the Cosmic Fig Tree that they wanted. As always, the tree granted them what they wished for, along with the built-in opposite. With sex came jealousies, with fame came isolation, with money came worries and with power came palace intrigues. And the net result was just more misery and suffering. Now the wish-granting power of the tree had become widely known and there was quite a throng of people wishing under the tree. And they were also equally miserable and suffered throughout their lives as they wished and wished.

By and by, as they became old men and women, the children congregate under the tree once again to contemplate their spent lives. They are now in three main groups. The first group is the Cynics, who say, "This tree duped us throughout our lives and made us miserable. It has all been a hoax and a farce."

They were fools, for they had learnt nothing.

The second group, the Wise-guys, say, "We must have been wishing for all the wrong things throughout our lives. We will wish again for all the right things to make us happy."

They were bigger fools, for they had learnt less than nothing.

The third group, the Death Wishers were the most foolish of the lot. They say, "This tree ruined our lives and made us miserable. We wish we were dead." And the obliging tree grants them their death wish, but they are immediately reborn underneath the same tree, for the tree always grants wishes in dualities.

Meanwhile a lame child had been watching all this from inside the window of the hut. He had wanted to wish for a good leg so that he could walk, but he was pushed away by the throngs of people crowding under the tree and he wasn't strong enough to get through. But as he watched from the window, he was awed by the spectacle of the tree. He saw his friends wanting sweets and clutching their stomachs with pain. He saw them grabbing their fancy toys and getting bored with them. He saw them wishing for sex, fame, money and power and suffering through their built-in opposites. As he witnessed the misery and suffering of the wishers, he began to understand the true nature of the brilliant cosmic swindle that was being enacted under the tree. With that understanding, he felt a well of compassion rise up from within him, not only for the wishers but also for all the creatures that were affected by the wishing. Through that gratuitous, all-encompassing compassion, wherein he sought no benefit from that compassion, the lame child lost his desire to wish and ensured his lasting happiness. He had sliced that fig tree with detachment. He had stepped outside the orbit of Karma and had done the pure act of kindness, Nishikama Karma. He was, without doubt, the happiest of the lot.

The Cosmic Fig Tree is, of course, the unsentimental universe and the wishes that it grants usually require creating some form of "order" in the universe. Such order is always accompanied by enough disorder so that the net disorder, or Entropy, in the universe constantly increases. This is known in physics as the Second Law of Thermodynamics. This is the source of the built-in opposites in the wishes that the Cosmic Fig Tree grants.

3.2 Two Pillars of Happiness

Thus, the Rig Veda revealed two of the main pillars of happiness, Compassion and Detachment, through this profound story. The Upanishads, which are commentaries on the Vedas, goes on to expound that both the wishing child, or the Participant (the Caterpillar), and the watching child, or the Witness (the Butterfly), are within each one of us. They are the built-in opposites within the make-up of any human being, and it is up to each of us to empower one over the other in our daily lives.

A note of clarification: while compassion is at the root of most religious traditions, both east and west, the English word, "Compassion," has slowly accumulated the connotations of "Tolerance" and "Pity," which results from a judgment of superiority on the part of the one feeling compassionate. Likewise, the English word "Detachment," has become conflated with "Aloofness" and "Apathy." This corruption of language stems from the corruption of our culture and it has proceeded to the point where I find that there are no appropriate words left to convey the original meanings. But part of the healing process requires us to reclaim the original meanings of these words.

I consider compassion to be the capacity to see clearly into the nature of suffering and to recognize that the suffering is not separate from ourselves, regardless of where it is occurring. There is absolutely no judgement involved in true compassion. But far too often, in organized religions, the circle of compassion is interpreted to be confined to just other human beings. The assumption seems to be that the circle of compassion needs to be enlarged step by step and humanity hasn't yet achieved the first step encircling fellow human beings. But as Albert Einstein said,

> "A human being is a part of the whole, called by us 'Universe,' a part limited in time and space. He experiences himself, his thoughts and feelings as something separated from the rest, a kind of optical delusion of his consciousness. This delusion is a kind of prison for us, restricting us to our personal desires and to affection for a few persons nearest to us. Our task must be to free ourselves from this prison by widening our circle of compassion to embrace all living creatures and the whole of Nature in its beauty. Nobody is able to achieve this completely, but the striving for such achievement is in itself a part of the liberation and a foundation for inner security."

Carbon Dharma

True compassion is all-encompassing.

Compassion goes beyond empathy. Jeremy Rifkin, in his book, "The Empathic Civilization[3]," argues that an upsurge in empathy will help humans overcome the resource bottlenecks that lie ahead in our civilizational endeavors. However, empathy is a natural trait common to all social species. It is customary for a dog to feel empathic suffering at the discomfort of another dog, or to feel empathic joy when the other dog is at play. However, it is not clear that any upsurge in empathy is in the offing for human beings, nor is it clear that any such upsurge will do humans much good. It is empathy that leads to the African concept of Ubuntu or instant Karma: when we hurt someone, we hurt ourselves; when we help someone, we help ourselves. It is because of Ubuntu that a good deed is its own reward and a bad deed, its own punishment.

Compassion occurs when we not only empathize with the suffering of ourselves or another, but also feel compelled to act to alleviate that suffering.

Detachment, on the other hand, is about the attainment of freedom or liberty. When neither praise nor criticism affects us, then we are truly detached and are free to pursue our autonomous path, to act without fear or favor, to dance our dance. It is boundless compassion accompanied by detachment that is at the root of true happiness or "Ananda" (Bliss) as the Rig Veda revealed in the Cosmic Fig Tree story. Such true happiness results from the absence of suffering within us while the kind of "happiness" that modern commercial interests promote is really pleasure, the opposite of pain.

Pleasure is fleeting, while happiness endures.

As Einstein noted, Man is, at one and the same time, a solitary, individual being as well as a social being[4]. Detachment, or true freedom. meets the individual needs of Man, while compassion meets the social needs.

Compassion without detachment will only lead to empathic torture as we take on our suffering and the suffering of others without the barrier of detachment. And compassion cannot be compartmentalized. It must shine on all creation without discrimination. For Ubuntu makes it next to impossible for a human being to wring the neck of a chicken and then turn around and be compassionate to another being.

Buddhism emphasizes detachment as the source of happiness. The Buddha famously observed that, "the world is full of suffering; the root of suffering is attachments; and the uprooting of suffering is through the dropping of attachments." But detachment without compassion may lead to solitary enlightenment, with not much benefit for the world at large.

How do we cultivate compassion with detachment? The Cosmic Fig Tree story tells us that in order to become compassionate and detached, we must learn to empower the lame child, the Witness, within us over the wishing child, or the Participant. As Anthony DeMello[5] and numerous authors before him have pointed out, mindfulness is the only tool we need to achieve that. But mindfulness is just a tool. It can be wielded by a brain surgeon as well as a burglar. Mindfulness does not promote happiness or well-being by itself, as the burglar has to be truly mindful to carry out his act, but he is most likely experiencing much trepidation as he goes about collecting his loot. The burglar certainly isn't happy when every little noise makes him jumpy. Therefore, while we must practice mindfulness to actively cultivate compassion and detachment and thus become happy, we won't automatically become detached, compassionate and happy, if we are simply mindful and in the moment as the example of the burglar illustrates.

Right action or Dharma is that which originates from such a state of boundless compassion with detachment. It is what the Witness within us, the Butterfly, would do if she were empowered. But our modern industrial culture is all about the wishing under the Cosmic Fig Tree. It is entirely geared for the Participant, the Caterpillar, who is wishing under the delusion that he is pursuing happiness through the fulfillment of desires but suffering through the bitter after-taste of disillusionment, while billions of people, animals and all of Nature become collateral damage.

3.3 The Addiction of Wishing

Wishing is an addiction, plain and simple. The desires which provoke the wishing are never ending, especially when a ubiquitous media is geared to stoke those desires. It is standard practice among modern sales professionals to use the AIDA concept[6], Attention, Interest, Desire, Action: attract the Attention of the potential consumer, invoke Interest, establish Desire and finally facilitate the Action of purchasing. We fall for this manipulation mainly because consumption has been promoted as the means to establish our social circle and then to define our place in it. For the Miglets, having been raised with cable television and the internet, this promotion of consumption began from their infancy onwards.

In general, the wishing becomes increasingly frenetic as the wealth possessed by the wisher increases. Society builds goods, exotic goods, with fancy artwork, using thousands of hours of human labor and tons of input from Nature, for one-time use for the rich. I was at a charity auction to support a non-profit and an incredible, hand-crafted, Indian silk dress was up for a bid. The bidding was fierce until there were just two women raising their placards. Both the women were at our adjoining table that was sponsored by the same rich couple and they obviously knew each other. Yet the women kept raising the stakes until the bid reached $5000 and the woman sitting next to me stopped raising her placard. The other woman won the dress.

I turned to my neighbor and said, "You know that your friend has the dress. You can always borrow it from her and wear it whenever you want."

She replied haughtily, "She wears it once and then that dress is finished! It will never be worn again by anyone!" It turns out that in her social circle, dresses are worn precisely once in public!

Later, I asked a reputed economist how can this continue forever with billions of people wishing for ever more one-time use goods, ad infinitum. And he replied: "Why not? With productivity improvement and price signals, infinite growth is possible." And he proceeded to use Moore's law in semiconductors as an example of productivity improvement that can drive the growth in economic activity, ad infinitum. Gordon Moore, a co-founder of Intel Corporation, famously said that the number of transistors that can be placed inexpensively in a semiconductor would double roughly every 2 years[7]. While Moore's law has held true to date, having been in the trenches of the semiconductor revolution, I believe that the economist is sadly mistaken in his assertion that Moore's law can continue for ever. For example, if Moore's law continues for the next 300 years, there would be more transistors on a single semiconductor device than there are atoms in the solar system, which is absurd. Nevertheless, even if we continue to build tinier and tinier devices with more and more functionality that we replace at a faster and faster pace until we literally run out of atoms, what is this being done for, except to fuel addictions?

Therefore, wishing is an addiction. This is why billionaires with more money than they can ever spend are trudging daily to work to wish for more money. Even if that means building and expanding businesses and enterprises that pollute and destroy Life on earth.

Addictions are hard to break.

During my college years, I took up smoking much to my mother's distress. She reminded me constantly that I was a sickly, asthmatic infant and that she had to massage my chest with warm, medicinal oil every day for the first two years of my life to help me breathe normally. Therefore, my smoking habit was truly a stupid move on my part, but from the moment I dragged on that first cigarette, I was hooked. And from the moment she found out about my smoking, Amma made it her mission to get me to stop. I tried to quit several times using nicotine patches, nicotine gum and even hypnosis, but couldn't. And during every telephone call and in every letter from India to the US, Amma pleaded with me to quit.

In February of 1997, Amma died in her sleep peacefully in her 60th year.

In March of 1997, I quit my smoking habit for good. I was devastated that I couldn't fulfill my mother's one simple request when she was alive and I was determined to quit. This time, it wasn't Amma's nagging that was making me go through the motions of quitting, but an irresistible force from within the depths of my being that was compelling me to quit. All the propaganda, advertising and chemical machinations of the tobacco purveyors were no match for this inner force. I quit cold turkey, even before understanding most of the things that I'm writing about in this book.

Human behavioral change occurs in one of two ways: through manipulation by external forces or through inspiration from within. It is change of the latter kind that is enduring. The former changes come from coercion or fear and can be reversed when the coercion stops, while the latter changes come from love. But often times, we need an eye-opening event to trigger that inspiration, to wake us up from our habitual stupor.

For the past few decades, Nature has been sending us distress signals that our addictive wishing under the Cosmic Fig Tree cannot continue. When a population of 100,000 tigers gets diminished by 97% within a century, the Holocaust that we are perpetrating on Life through our addiction is unmistakable. Yet we continue the wishing furiously, while inventing newer and fancier technologies to stoke the desires of the wishers. Indeed, most of Google's considerable annual revenues come from advertising related to searches, which is the pin-point targeted pushing of this addiction. Facebook promises to further refine the targeted advertising by using all our social history and our friend circles to select the products for our potential consumption. This is personalized AIDA on steroids.

Should we not quit this addiction of wishing, our Caterpillar lifestyles, before our common mother, Earth, dies? Or is there a way to continue our Caterpillar lifestyles while using alternate energy sources to fuel the Cosmic Fig Tree's benevolence? If the latter, can this be done while raising the wishing opportunities for the nearly six billion other human beings who aren't indulging in the tree's bounty to date? And can this be done while preserving the biodiversity wealth of the planet? These are the questions of our times.

Fig. 3. *The cancer underlying the fever: our destruction of Nature. Over 30 million acres of forests are being cut down each year, mainly to grow livestock feed and biofuel crops.*

4. The Really Inconvenient Truth

"If we save the living environment, the biodiversity that we have left today, then we will also automatically save the physical environment. If we only save the physical environment, then we will ultimately lose both" - Edward O. Wilson.

Former Vice President Al Gore changed my life. It was his slide show that stopped me dead in my technical career track to change course and devote the rest of my life to environmental causes. It was a cold evening in December of 2005 when I came to a warm home in New Jersey, switched on Link TV[1], a public service channel on our satellite dish, and watched Mr. Gore present his slide show before a small audience in San Francisco. This was a few months before his Oscar winning documentary made such a splash worldwide, and someone had captured it with a hand-held camera. Despite the simple production of the video, I was so shocked by what he was showing that I told my wife that if even half of that was true, we were wasting our time at our start-up trying to create new technology and ultimately, accumulate more wealth for our future. What was the use of leaving our children a pile of money when the planet itself was going to be desolate and unfit to live in if we don't solve the climate crisis? A few months later, after poring through the environmental and scientific literature, I came to the conclusion that the situation was actually far worse than what was portrayed in the slide show. Then we closed our start-up and I wrote to Mr. Gore offering to help in any way I can. This is how, in December of 2006, I became one of a few thousand trained presenters of his slide show and I went around presenting it in New Jersey and India over the next year on behalf of the Climate Project[2], a non-profit that Mr. Gore founded to raise awareness on the climate crisis.

The Oscar winning documentary, "An Inconvenient Truth[3]," (AIT) based on Mr. Gore's slide show was released in mid-2006 and it raised the profile of the climate crisis in the world at large and won Mr. Gore the Nobel Peace Prize for 2007, jointly with the scientists at the UN Intergovernmental Panel on Climate Change[4] (IPCC). Through the skillful use of animations, movies and imagery, AIT conveyed the scientific view that the world was experiencing increasing fires, floods, droughts, hurricanes, tornadoes, ice-melts, loss of arctic ice and sea level rise due to the accumulating greenhouse gases in the atmosphere, especially carbon dioxide (CO2), spewed out mainly by the fossil-fuel based energy infrastructure of the industrial world. The increasing concentration of greenhouse gases was trapping more of the solar energy falling on Earth and heating it up rapidly. But AIT left the impression that we can resolve these issues if we only had the will and the wherewithal to deploy clean, renewable energy solutions for the world's ever-increasing energy needs. Today, I believe that Mr. Gore was more on the mark when he wrote in his book, "Earth in the Balance[5]" in 1993, that "The more deeply I search for the roots of our environmental crisis, the more I am convinced that it is an outer manifestation of an inner crisis that is spiritual." And such a fundamental, inner crisis cannot be fully solved merely with external technological deployments. Nevertheless, for a while after being trained by Mr. Gore, I conveyed the impression to my audiences that technological solutions do exist for our predicament.

I hope that they will accept my apologies now.

Both Mr. Gore's slide show and the IPCC reports are mainly conservative and portray a best-case projection because the process of achieving scientific and political consensus tends to gloss over and under-report extreme future scenarios. The Earth's climate system is a discontinuous, chaotic process and all indications are that the discontinuities add to our discomfort, while most scientific projections assume a continuous model for climate change. It is hard to predict when certain tipping points that can catastrophically accelerate climate change occur and the IPCC analyses assume that these tipping points are not reached. In particular, the IPCC report is hammered out with scientists and political representatives from 192 nations reaching consensus over the content. Can you imagine the Saudi Arabian delegation, for instance, agreeing to scientific data on climate change that is not iron-clad, since the economic health of its country and the political health of its leadership rests on the world's continued dependence on fossil fuels?

During the Climate Project training, Mr. Gore told us to be careful not to deplete the "Hope Budget" of our audience. The key was to keep the audiences convinced that their present lifestyles will be preserved as the climate crisis gets solved. They must leave with the impression that modern industrial civilization as we know it will continue, but with wind turbines and electric cars replacing coal power plants and gas guzzling SUVs.

But audiences, above all, deserve to be told the truth as the speakers see it. A talk by my colleague, Dan Miller, entitled "A Really Inconvenient Truth[6]," where he lays out the actual worst-case, nominal and best-case scenarios that would be unfolding due to climate change, is one of the most widely viewed presentations on Fora.tv. But Dan's presentation deals only with climate change and doesn't consider the other environmental crises that are unfolding on the planet, the nitrogen cycle disruption, chemical pollution, radiation leaks, species extinctions, etc. Modern industrial civilization is already on life-support as it has come to depend upon extreme sources of raw materials with all the attendant risks of mega disasters. Two such mega disasters occurred in the past two years with the British Petroleum (BP) Gulf Oil Spill of 2010[7] and the Fukushima Daiichi nuclear disaster of 2011[8] disrupting millions of human lives in the US and Japan. The repercussions of the spilled oil in the Gulf of Mexico and the radiation leaks from Fukushima will be felt for decades to come as these pollutants are processed through the biosphere of the planet over time. And so will the repercussions from future accidents at the thousands of other extreme technology sites that modern industrial civilization has deployed throughout the world. Fukushima Daiichi and the BP Oil Spill were "explosive" events that caught peoples' attentions, while the steady release of greenhouse gases and pollutants from coal power plants, automobile exhausts and animal agriculture are drip, drip, drip events that don't trigger the threat response from human beings, even if they pack a bigger punch on a long-term scale.

In a tacit admission that AIT and the Climate Project have failed to sway America, Mr. Gore recently renamed his non-profit as "The Climate Reality Project[9]" and delivered a shorter, punchier slide show to turn things around. However, the term, "Climate change" no longer invokes a sense of urgency in the American public. The opposition has cleverly co-opted this phrase by pointing out that the climate is always changing. So what's the problem? The rise in temperature due to climate change does not seem to bother Americans so much. Their air-conditioners may have to work a little harder if the temperature gets too hot, but as long as the utility companies continue to supply electricity into the homes, what's the problem?

The sea level rise that Mr. Gore showed in AIT seemed speculative, too far out and too gradual to get people to act now. Americans can always pick up and move if the ocean starts lapping at their door step. Or if they live in a city, they can always build a retaining wall. Isn't the Netherlands already way below sea level and haven't the Dutch managed to keep the sea out of their land? So what's the problem?

The opposition likes to portray AIT as scare mongering. More intense hurricanes, more heavy downpours, more floods, more droughts, all more of the same natural forces that people have lived with for ever. So what if the minimum volume of ice in the Arctic has diminished by more than half in the past 5 years alone[10]? Even this Arctic melting may not be so dire for the polar bears after all, as they seem to be resourceful enough to adapt their food habits. Besides, the Arctic melting opens a Northern route for shipping and allows the erection of oil drilling platforms in the Arctic to meet the rising energy demands of the world. If the Arctic becomes ice-free in summer in 20 years as the scientists are now predicting, then that's surely great for world commerce as New York and London become so much closer to Tokyo and Beijing for shipping traffic.

So what's the problem? At the opening of a recent shale-gas conference[11], Karl Rove, the keynote speaker said, "Climate is gone!" He assured the attendees that they won't need to worry that the new Congress will consider any legislation on the environmentally destructive practice of "fracking" to extract natural gas from formations such as the Marcellus Shale in Pennsylvania. There were cheers all around. On the policy front, the new US Congress seems to want the country to rapidly backpedal on all its environmental obligations, including long standing ones stemming from the Clean Air Act and the Clean Water Act of the 1970s. It has now added exemptions to these environmental laws to promote fracking and thus facilitate the unconventional extraction of oil from the Tar Sands of Alberta, Canada[12].

In hindsight, even if the policy prescription in AIT had been widely adopted, it wouldn't have succeeded because: 1) AIT focused on a symptom and not the disease, 2) it mis-diagnosed the root cause and 3) it glossed over the deep cultural changes required to address the root cause.

4.1 The Core Problem

In 2010, the late humanitarian and scientist, Dr. Stephen Schneider of Stanford University, was addressing a roomful of skeptics on climate change in Australia[13]. He was asked the inevitable question, "Since CO2 is good for plant growth, what's the problem with humans emitting CO2 as part of our industrial activities?" Dr. Schneider patiently explained that yes, some plants grow faster with higher atmospheric concentrations of CO2, but other plants don't. Then the faster growing plants crowd out the slower growing ones and kill them off, thereby upsetting the balance in the ecosystem and possibly triggering an ecosystem collapse.

The true reason that climate change is a crisis is that it is happening so fast that ecosystems are having a hard time adapting to the environmental changes around them and are dying off. It is hard for trees to pick up their roots and move because New Jersey has now acquired the climate of Virginia. At the current pace of climate change, within a few decades, New Jersey is expected to experience the recent climate of North Carolina making it that much harder for ecosystems to adapt. Vast swathes of forests in Rocky Mountain National Park in Colorado already contain dead pine trees due to bark beetle infestation. British Columbia in

Canada is projected to lose almost all of its forests over the next five years for the same reason. The bark beetles migrated to these regions in response to the change in climate, and pine trees have not yet evolved any natural defense against them.

The ice core data from Greenland and the Antarctic reveal that the rate of change of CO2 due to human activities today is orders of magnitude faster than that experienced through natural cycles. A 30ppm (parts per million) increase in CO2 occurs in the ice core data over a 1000 year period, while that same increase occurs in a span of 17 years at the present time[14]. As a result, NASA scientists estimate that the rate of movement of isotherms (lines of constant average temperature) has also been an order of magnitude faster than the response rate of ecosystems[15]. A Nature Geosciences study in 2010 found that the ocean is also acidifying an order of magnitude faster than when a mass extinction of species occurred 55 million years ago during the Paleocene-Eocene Thermal Maximum (PETM)[16]. And ocean acidification occurs when part of the CO2 increase in the atmosphere is absorbed by the water in the ocean to form carbonic acid. Rapid acidification leads to the death of marine species that need to form shells and with their demise, the food web in the ocean tears apart and disintegrates.

Therefore, it is Life that is under siege from climate change. The German scientist, Hans Joachim Schellnhuber, recently asserted that the proper analogy for the Earth's projected temperature rise is a corresponding rise in human body temperature due to a fever[17]. The normal human body temperature is 37°C, and at higher temperatures, the human body reacts increasingly violently. Here's a list compiled from Wikipedia[18]:

• +1^{O}C: leads to sweating, feeling very uncomfortable, and feeling slightly hungry.

• +2^{O}C: leads to severe sweating, flushed and very red, with increased heart rate and breathlessness. There may be exhaustion accompanying this. Children and people with epilepsy may be very likely to get convulsions at this point.

• +3°C: leads to fainting, dehydration, weakness, vomiting, headache and dizziness as well as profuse sweating. The fever starts to be life-threatening.

• +4°C: leads to fainting, vomiting, severe headache, dizziness, confusion, hallucinations, delirium and drowsiness. There may also be palpitations and breathlessness. This should be treated as a medical emergency.

• +5°C: leads to coma, severe delirium, vomiting, and convulsions. Blood pressure may be high or low and heart rate will be very fast. Subject may turn pale or remain flushed and red.

• +6°C: leads to death or severe brain damage through cardiovascular or respiratory collapse, continuous convulsions and shock.

Here's a list compiled from the British journalist, Mark Lynas's book, "Six Degrees[19]," for an increase in the Earth's average temperature:

• +1°C: leads to mega drought in the American west from California to the Great Plains, rivaling the dust bowl of the 1930s, depleting the agricultural productivity of the land. Europe, Africa, Asia and Australia will also face similar challenges as rainfall patterns are altered.

• +2°C: leads to drought-stricken cities in China and food emergencies throughout the world. Life in the ocean collapses leading to famine among dependent populations. Corals die. Europe wilts as it experiences the present climate of North Africa.

• +3°C: leads to a persistent super El Nino, the drying out and burning of the Amazon, the drying out of the Indus and Colorado rivers, extreme hurricanes, rapid sea level rise, famine and starvation throughout the world.

• +4°C: leads to Antarctic ice melts, eventual sea level rise of over 50 meters and possibly even an ice-free Earth, deluging not just our cities, but also our nuclear power plants and their radioactive waste dumps.

- +5°C: leads to an entirely unrecognizable planet Earth, with rain forests all burned up and humans herded into shrinking zones of habitability by the twin crises of drought and flood.

- +6°C: leads to near certain, rapid mass extinction, possibly "Game Over" for humans.

At present, the Earth has experienced a 0.8^{O}C (1.4^{O}F) rise in average temperature, which is equivalent to a human being running a 100°F fever. Now imagine going to a doctor with a persistent 100°F fever and the doctor tells you not to worry about it because such body temperatures are normal in cattle[20]! This is precisely the analogy when contrarians point out that the Earth used to have higher average temperatures several million years ago. Besides, notice how this year's mega drought in Texas and in the American southwest resembles the above projections? Nevertheless, scientists have been reluctant to attribute the 2011 drought to climate change partly because of the inherent conservatism of their profession and mainly because a single event does not imply a long-term trend. But the preponderance of extreme events recently, the Russian heat wave, the Pakistani floods, the record tornado season and the drought in the American southwest all do point to the Earth experiencing a fever that is consistent with the above expectations. In addition, the extremely conservative IPCC is projecting anywhere from 1.7^{O}C to 5.8^{O}C increase in the Earth's average temperature by 2100 under various scenarios. An MIT study projects a 50% chance that the average temperature will increase by over 5°C by 2100 in their business-as-usual scenario[21]. The UK Met Office has a worst-case prediction of a 4°C rise in temperature by 2060, due to carbon cycle feedbacks[22].

And that would be within the lifetime of the Miglets.

Complex life dies out with rapidly changing temperatures on Earth because ecosystems don't have access to climate controlled environments as humans do. It is true that, given cheap access to sufficient energy, humans can crank up their air-conditioners and ride out the effects of climate change in the confines of their homes and automobiles, but the same doesn't apply to wildlife, marine life, crops and other sources of the food that we eat. Nevertheless, at present, climate change is not yet the major reason for the loss of Life that is ongoing on the planet. Human consumption due to our industrial, Caterpillar culture, tops the list. Complex Life, as we know it, is dying off on the planet, mainly because we're killing it and literally eating it up, directly and indirectly. Two-thirds of the ice-free land area of the planet has already been transformed for human use such as for agriculture, cities and livestock production, with less than one-third left for wildlife. Since three-fourths of humanity are still clambering up the development ladder and are yet to partake of the largess of the planet as the top quartile have been doing, they are increasing pressure on that one-third of the land currently used by wildlife. As a result, the Holocaust on complex Life is well and truly underway. Even the Western environmental movements of the sixties and seventies may have perversely accelerated this Holocaust by distancing the average Western consumer from the direct consequences of his consumption.

In her book, "The World is Blue[23]," Dr. Sylvia Earle, the National Geographic Explorer-in-Residence recalls an encounter with Kuzno Shima, the head of the Japanese delegation to the International Whaling Commission during the 90s. He challenged Dr. Earle with questions such as, "Americans eat beef, right? What's the difference between eating steak from a cow and eating whale meat?" Dr. Earle responded earnestly contrasting the agricultural production of cows with the wild life of a whale and arguing that there were a billion plus cows on the planet whereas there were only a few thousand whales left. Shima listened patiently but was not moved, which Dr. Earle couldn't fathom. As I read this passage in the book, it occurred to me that Shima was on to something and that Dr. Earle was missing his point. After all, to raise a billion plus cows and other livestock on the planet, humans have appropriated nearly one-third of the ice-free land area of the planet, displacing numerous other species and decimating their numbers. While

Americans may not have eaten all the mountain lions, the Indians may not have eaten all the tigers and the Chinese may not have eaten all the giant Pandas, directly, they all might as well have done so. They certainly caused the habitat losses that have resulted in the near extinction of these magnificent animals through their appetites for beef, milk and pork, respectively. It is these second-order effects on Life of our ever-increasing ecological footprints on the planet that even great scientists such as Dr. Earle have failed to grasp and articulate.

Let us take, for instance, milk consumption in India. Most Hindus venerate the cow and do not eat beef, but they drink milk and eat cheese. In Western countries, the dairy cow is ruthlessly chopped up into hamburgers as soon as its milk production declines at the age of four[24], while the typical Indian cow lives out to an old age of 20 plus years, grazing on forest and other pasture land. This grazing reduces food for the sambhar deer and other wild ruminants which decline in population, putting a downward pressure on the tiger population. And the whole ecosystem suffers. This is why I realized that if I drink milk, then I must be prepared to eat the beef when the dairy cow ceases to be productive and I must be prepared to eat the veal from the male calves of cows, in order to optimize my ecological impact. Otherwise, there would be an order of magnitude more cows alive for a given level of milk production, which does happen to be true in India. And as I drink milk in India, I'm effectively eating the tiger and the sambhar deer, etc. Once this realization dawned, I became vegan instantly. And within a couple of months after turning vegan, I became lactose-intolerant as my body adjusted to my new dietary intake and my Gastro-Intestinal (GI) tract elongated according to my nephew, Arun, who's a Gastro-enterologist. Now, I have a built-in mechanism that makes me feel awful if I accidentally consume any animal product, because that animal food becomes rancid when traversing my lengthier GI tract.

The American environmental movement of the sixties and seventies resulted in a slew of important legislation such as the Clean Air Act, the Clean Water Act and the Endangered Species Act. While the air, water and endangered species in America were better protected as a result of these measures, they also caused the outsourcing of manufacturing operations to developing countries where such environmental measures were non-existent. In the meantime, American consumer desires continued to escalate so that the net pollution on the global environment kept increasing, but far from where the consumers lived. When I asked a reputed environmentalist whether these Acts were really such good ideas if they simply outsourced the destruction of the environment to Third World countries, he responded with "none of it is perfect, but I've seen no golden bullet for changing consumer desires so far." As we protect land in the US to save one species or the other, we simply make life harder for numerous species, for example, in the Amazon, in what is, at best, a questionable tradeoff.

Therefore, focusing on climate change as the primary problem to address is like focusing on the fever resulting from a rapidly spreading cancer in our body. While we need to keep a healthy stock of aspirin to douse the fever as it gets serious, it is more important to undergo the cancer treatment. And it is even more important to stop ingesting carcinogens on a daily basis.

AIT focused on the Earth's fever, but glossed over the cancer that is our Caterpillar culture. One of my fellow presenters at the Climate Reality Project (TCRP) wrote me to the effect, "You are trying to save 'climate through species,' while we are trying to save 'species through climate.' In the long run, our approach is better because people don't really care about biodiversity." And thus, he highlighted the true tragedy of my generation, that we are standing by idly with folded hands while the biodiversity of the planet is ravaged by our Caterpillar culture, thinking that the majority of people don't care that it is being ravaged. As Albert Einstein said, "The world will not be destroyed by those who do evil, but by those who watch them without doing anything."

Perhaps, it is the steadiness of the biodiversity decline that lulls us into this apathy. When I was growing up in India, I used to participate in a ritual on each anniversary of my grandfather's death, when my mother would make a feast and arrange the food

items on a banana leaf outside, while my siblings and I took turns
to call the birds. Usually, we would attract a flock of crows to come
by and feed on the offerings, signifying that our grandfather had
accepted our prayers and remembrance. After my parents passed
away, I happened to be in India during one of their death
anniversaries and I asked my sister why we are no longer practicing
this ritual. She told me that there were very few birds left in the
cities of India and that it would be devastating if the offerings were
not accepted. Therefore, this ritual is no longer in common practice
today.

Scientists have long recognized that if we care to look up in the sky
and count the number and diversity of birds that are flying around,
we would get a good indication of the extinction that is already
under way in our local area[25]. When it comes to energy and
minerals, humans have barely scratched the surface of the Earth. I
believe that humans are perfectly capable of developing more
powerful technologies to access the "inaccessible" sources of these
necessary ingredients for industrial civilization, if need be. On this,
I agree with the cornucopian economists that human ingenuity will
prevail when it comes to material resources, with the caveat that as
we deploy more extreme technologies to access those resources, we
will be undertaking more extreme risks as well. But there is no
hidden reservoir of keystone species to revive the ecosystems that
we destroy. And it is hard to rebuild ecosystems when we don't
know 95% of the species that were in the ecosystem in the first
place. Despite our technological prowess, humans are not very
good at rebuilding complex systems as one disaster after another
has illustrated, starting with Hurricane Katrina and continuing on
through to Fukushima. How smart are we really when the best
engineering minds in the world can only stand around and watch as
nuclear reactors melt down and irradiate vast swathes of our
planet?

4.2 The Root Cause

Since AIT focused on the symptom, it identified the fossil fuels
that we burn to drive our industrial civilization as the root cause of
our predicament. But if the core problem is truly the extinction of

Life, then the root cause is human behavior, our Caterpillar culture, specifically our culture of violence towards Nature, our neglect, abuse and consumption of Life and our thoughtless meddling with the biosphere of the planet. And climate change due to our fossil fuel burning is still a small part of the reason for the carnage around us, though a growing one. If we truly viewed Life on Earth as precious, to be preserved and celebrated for our own enlightened self-interest since no other planet in our galactic neighborhood has good life-support systems even if we managed to get there en masse, then we wouldn't be so mistreating Life as we do now. Sri Aurobindo, the famous 20th century Indian philosopher, said,

"Humanity's true moral test, its fundamental test, consists of its attitude towards those who are at its mercy - animals. And in this respect, humankind has suffered a fundamental debacle, a debacle so fundamental that all others stem from it."

We don't have to be experts in paleoclimatology or in the computer simulation of fluid dynamics to understand the non-sustainability of our mistreatment of Life. Just simple arithmetic would do. In a letter to Mr. Gore that I wrote with a few of my Climate Project colleagues, we explained, "The 2006 United Nations FAO report, 'Livestock's Long Shadow[26],' estimated that humans use nearly one-third of the ice-free land area of the planet, or nearly twelve billion acres, for livestock production. According to many scholars, such as Professor David Pimentel of Cornell University, just over one-fourth of humanity consumes most of these livestock products[27]. To put that number in perspective, the Global Biodiversity Assessment of 1995[28] estimated that the Earth can sustainably support just one billion human beings at American levels of consumption." We concluded that the enormous footprint of humans on Life through our consumption of meat and other animal products is an Elephant in the Room that needed to be addressed in our presentations, especially to American audiences. Nevertheless, in an interview with the host Larry King on Cable News Network (CNN), Mr. Gore went on to publicly admit that he continues to eat meat, because he "likes the taste[29]."

Later, I was attending a speech by Jim Hansen, the eminent climate scientist from NASA Goddard Institute of Space Studies at the Chabot Space Center in Oakland, CA. His speech was arranged as part of his book tour to introduce "Storms of my Grandchildren[30]," an expertly crafted case for action on climate change mitigation. Dr. Hansen's reason for writing this book was that his grandchildren might otherwise say that "While Opa knew that climate change was serious, he didn't do anything about it." However, during the Q&A session after his talk, Dr. Hansen also admitted that he continues to eat beef and other meats.

What is troubling about these admissions is that it is now widely known that the embodied energy that is used in the animal agriculture systems of the planet far exceeds the energy used for transportation and even for the heating and cooling of buildings. Animal agriculture is the modern practice of raising animals as if they were mono-cultured crops, in vast factories, to be slaughtered and used as commodities for human consumption.

In general, any time we defy Nature and replace native ecosystems with mono-cultured crops, we pay a price by creating an imbalance in the Earth's carbon cycle for one simple reason: native ecosystems are known to maximize the carbon stored on land, the carbon sequestration, in any given region[31]. Further, this imbalance gets amplified by an order of magnitude when we replace native ecosystems with animal agriculture where we feed the factory-grown animals with mono-cultured crops such as corn and soybeans, for most of what we feed the animals turns into just manure. Ultimately, climate change is a direct consequence of the imbalance we're creating in the carbon cycle through human activities, where we're essentially transferring naturally sequestered, land-based, carbon into the atmosphere in the form of gaseous carbon such as CO_2 or methane, thereby amplifying the greenhouse effect.

We got an inkling of the impact of animal agriculture when the UN published its Livestock and Climate Change report in 2006, where the livestock sector was calculated to be contributing 50% more greenhouse gas emissions (18%) than the entire transportation sector of the world (12%). Later, in a 2009 Worldwatch Institute report, two UN Environmental Assessment (EA) specialists, Robert Goodland and Jeff Anhang, pointed out that the 2006 UN report failed to take into account the carbon cycle imbalances caused by the conversion of forests to livestock pasture lands. They came up with an estimate that the livestock sector was responsible for 51% of world greenhouse gas emissions[32] but their calculations based on the breathing contribution of livestock were not widely accepted. However, in 2010[33], Prof. Danny Harvey of the University of Toronto in Canada, building upon the thesis work of Stefan Wirsenius from the Goteborg University in Sweden from 2000[34], calculated that the average human being is consuming more energy in food than in fuel and shelter combined, when we take into account the embedded plant-based energy input to the animal agriculture systems. And the main reason is that animal agriculture is so inefficient that, on an average, it requires 100 Joules of embedded plant-based energy to produce less than 4 Joules worth of animal foods such as eggs, dairy and meat for human consumption.

But if such simple arithmetic is hard to grasp for eminent scientists and thought leaders, should anyone be surprised that a substantial segment of the American public is being duped over the complexities of climate science by a well funded campaign of deception? Besides we don't really have to convert all the ice-free land on earth to livestock production and eat up all the fish in the ocean before triggering ecosystems collapses; perhaps, our current consumption and pollution level itself is sufficient to trigger such collapses as the species extinction rate is already 100-1000 times the background rate. Just as cancer doesn't have to consume all the organs in the human body before the human dies. When ecosystems collapse, they most certainly transfer their sequestered carbon into the atmosphere, thus creating a reinforcing feedback through climate change.

The trouble is that we view other life-forms to be inferior to us and to be dominated over and this has become a routine part of our industrial, Caterpillar culture. We don't even think about it, but just do it. The biologist Richard Dawkins once wrote[35], "Science boosts its claim to truth by its spectacular ability to make matter and energy jump through hoops on command." In that one sentence, he captured the fundamental axiom of the culture of industrial civilization, that of domination over and separation from Nature.

Domination over Nature is a fundamental axiom in the culture of modern industrial civilization that drives the separation from Nature as it is not possible to dominate something without becoming alienated from it. And our "Domination" over Nature is absurd, considering how powerless we are in the face of earthquakes, volcanoes, tsunamis, hurricanes, floods, droughts and other calamities. However, with our superior tools and technologies, we certainly have the prowess to dominate over other species, to capture, imprison and kill them at will, to commit a Holocaust on them.

The Nazis attempted to eradicate the "lesser people" of the planet. My generation is blindly eradicating the "lesser species" of the planet. History has not been kind to the Nazis or to the silent witnesses of the Nazi holocaust. History will most likely not be kind to my generation either even though it is merely continuing the inherited practices of previous generations, but with exponentially growing potency.

But it is hard to become aware of the holocaust when we are in the midst of it. It isn't just the factory farms and the slaughterhouses that we studiously ignore during our supermarket purchases that constitute the holocaust. We also pour pesticides to protect our crops, which cause insects to die as intended, but which also cause birds to die that fed on these insects and which cause fish to die when the pesticides run off into the streams and rivers. We use fertilizers, fungicides and herbicides on our farms and on our lawns to improve monoculture crop yields, but that deadens the soil and leaches poison into our environment. We ingest and excrete pharmaceuticals and pour other industrial and household chemicals into the environment so that most of our fresh water bodies are polluted and bereft of complex Life. As a result, we have to carefully treat our fresh water sources before drinking the water. A common water treatment system for human consumption is "a multi-step process involving micron pre-filtration, reverse osmosis separation, double ultraviolet sterilization and post carbon polish," to quote from the side of the vending machine that supplies our drinking water at home. But who is supplying such multi-step processed water to wildlife? Is wildlife not susceptible to cancer and other diseases that we're trying to avoid among the human population with our water treatment processes? I know that animals can contract cancer as our older dog, Midnight, is suffering from a terminal version of it at the moment. I also know that this is a common occurrence since there are more pet oncologists than human oncologists in our vicinity.

Nevertheless, as Jeremy Rifkin has pointed out repeatedly, not one of the 192 world leaders meeting at the UN nor Mr. Gore has highlighted our mistreatment of Life, even though it is one of the leading causes of climate change, well ahead of fossil-fuel use for transportation. It has been left to Mr. Rifkin and out-of-the-box thinkers such as the Professor from Harvard, Edward O. Wilson writing about biodiversity loss[36], or the journalist from NY Times, Mark Bittman, writing about the Western world's addiction to meat[37], to prod humanity to ponder in that direction. As Mr. Rifkin said, "How serious are we, if we don't even talk about meat consumption[38]?"

But denial is a standard response to addictions.

4.3 The Way Forward

Suppose that AIT's prescription of a rapid revamping of the energy infrastructure to renewable sources in a World War II style effort had been adopted. That would have rapidly eliminated the aerosols that are currently masking the true effects of the greenhouse gases in the atmosphere as the BBC documentary on "Global Dimming[39]" showed. Aerosols are microscopic particles such as sulphates and ash that both scatter sunlight directly and form the nucleus of water droplets that reflect sunlight back into space. As such, aerosols cool the planet and they are also the byproduct of our fossil fuel emissions. While aerosols last only 3-5 days in the atmosphere, the CO2 that we spew into the atmosphere from that same fossil fuel burning can last for centuries unless we deliberately regenerate forests or deploy other means to sequester the carbon back into the ground. If aerosols are suddenly eliminated from our atmosphere, the result would be an equally sudden spike in global temperatures across the planet, perhaps on the order of 1C or more. But can you imagine the political fallout, especially in the United States, if global temperatures spike up as renewable energy gets deployed rapidly on a vast scale as AIT prescribed? As it is, even the staid Economist magazine wrote regarding the climate disruption that has already occurred, "When reality is changing faster than theory suggests it should, a certain amount of nervousness is a reasonable response[40]."

On the other hand, forests sequester carbon and therefore, the regeneration of forests has the potential to reduce the CO2 buildup in the atmosphere. Joseph Canadell, a scientist at CSIRO, Australia's National Climate Research Center in Canberra said that "if you were to stop deforestation tomorrow, the world's established and regrowing forests would remove half of fossil fuel emissions[41]." But stopping deforestation and promoting afforestation would require addressing the number one cause of the deforestation on the planet, the ever increasing demand for animal foods and animal products among humans.

Instead AIT mainly focused on the conversion of the energy infrastructure of the world to clean, renewable sources as the panacea for Climate Change. It left the impression that there is little to no change required from the public at large, other than to muster the will to call their political representatives to act on legislation to support this conversion with appropriate top-down subsidies and policies. As such, AIT was based on the predicate that it isn't our Caterpillar culture, the addiction of wishing under the Cosmic Fig Tree that needs to change, but the fuel that powers the benevolence of the tree that needs to change.

However once we acknowledge that our behavior, our culture of violence towards Nature is the fundamental root cause of the environmental crises we face, then the public at large becomes the central actor in addressing the crises. All the catastrophic projections in AIT assume that we will continue to act as we have been doing regardless of the changed environment around us, that we will continue our inherited culture of violence, but somehow avoid the consequences by quickly deploying solar panels and wind turbines. This doesn't seem possible. Therefore the battle is really over the hearts and minds of human beings. This is the Kurukshetra from the Mahabharata of our times.

Therefore, human behavior, our over-consuming Caterpillar culture, the culture of violence towards Nature is the root cause, extinction is the disease and climate change is the symptom. Just as every good medical doctor knows, the fever is the symptom, not the disease by itself. Can you imagine going to a doctor with a cancer raging within you that has eaten half your vital organs and being told to take aspirin to douse the $100^{o}F$ fever that is a symptom of the cancer? Such a doctor would be considered incompetent. But that is the Really Inconvenient Truth that AIT glossed over.

And the way forward is to train every human being to see through the fear mongering and the incessant marketing of conspicuous consumption in our mass media, to practice mindfulness, cultivate compassion, detachment and patterns of conscious consumption and thereby, strive for a deeper sense of happiness than the Pursuit of Affluence that we are being steered towards. Our goal must be to create a just, caring world of abundance in which all people and all species have a chance to flourish.

Fig. 4. *The Elephant, a metaphor for the Truth.*

5. The Kurukshetra of our Times

"In the end, we conserve only what we love. We love only what we understand. We understand only what we are taught" - Baba Dioum.

I had a memorable childhood growing up in a Hindu Brahmin household with considerable exposure to the myths and fables from the great Epics and the Puranas of India. But while growing up, I never quite understood the significance of those stories and came to treat them with disdain. Being of a literal bent of mind, I couldn't understand how a ten-headed man, Ravana, could have existed at all, just to take an example. One of my school teachers was fond of quoting archaeological and physical evidence for the historical basis of some of the epics, which made it even worse for me to identify with them. He pointed out that there was evidence of a physical bridge between the southern tip of India and Sri Lanka, which to him, proved that the battle between Lord Rama and Ravana as described in the Ramayana must have truly happened. The rag tag army of animals and people headed by Lord Rama must have truly crossed over into Sri Lanka on foot to defeat Ravana and recapture Sita from his clutches. My teacher also said that archaeological excavations and astronomical signs indicated that the Battle of Kurukshetra, as described in the Mahabharata, must have actually occurred about 5100 years ago. In the Hindu LuniSolar Calendar, Kali Yuga started when the battle of Kurukshetra ended and we are currently in year 5112 KY[1], cementing his assertion. He concluded that the fantastic missiles and other weapons described in the Mahabharata constituted lost know-how from an advanced technological civilization that existed in India, which got destroyed in the immense conflagration of the Battle of Kurukshetra.

I concluded that he didn't know what he was talking about. Ten-headed men, seven-headed snakes and multi-limbed Gods and Goddesses were figments of the imagination and couldn't have actually existed as far as I was concerned.

I gravitated towards Science and Technology instead. It helped that my father worked in the British Council Library in India and as a result, I could read all the English scientific books that I wanted. In Independent India, successive governments had instituted a number of policies to redress the historical inequities of the caste system that had plagued Hindu society over the centuries. Therefore, as an upper-caste, Brahmin child in India, I had access to very few avenues for higher education that didn't involve a huge capitation fee (a form of bribe), a fee that my parents couldn't afford. Of these few avenues, the strictly merit-based admission scheme through the Joint Entrance Exam for the Indian Institutes of Technology[2] was the most attractive. That is how I became an Electrical Engineer and wound up in the US working on the hardware nuts and bolts of the Internet communications revolution.

5.1 The Symbolism of Idols[3]

I wish I could go back and relive my childhood. For a few decades, like the idiot in the Chinese proverb, I had been focusing on the wise man's finger and missed all the heavenly glories of the moon that he was pointing to. Like every good epic, both the Ramayana and Mahabharata were truly mixtures of historical events and allegoric fiction that conveyed the truth on so many levels. In the Ramayana, the ten-headed Ravana, whose heads magically grew back every time they were cut off, represents the never-ending nature of human desires, while Sita is the human spirit, the Atman, imprisoned by Ravana. Sita is rescued by Lord Rama, God himself, with the help of all the animals and birds of the world, by vanquishing these human desires. This is the story of the Cosmic Fig Tree all over again.

The numerous Gods and Goddesses of Hinduism are symbols representing various aspects of the path of Dharma. For instance, Lord Ganesha is depicted as a man with an elephant head symbolizing the sacredness of all Life forms. He represents ingenuity as he famously won a racing contest over his brother through his cleverness. In a dispute with his brother over a sweetmeat, their parents agreed to give the treat to the brother who was the fastest at circling the universe three times. While his brother proceeded to rush around the universe, Lord Ganesha simply circumambulated around his parents three times and won the contest. That is ingenuity.

Lord Ganesha is usually shown with four or more hands, clutching the symbols for the characteristics that his devotee aspires to. One of his hands is held in a gesture known as "Abhaya Mudra" denoting courage or fearlessness. In another hand, he holds an axe reminding the devotee to cut off all his attachments. In a third hand, he holds a rope intended to pull the devotee towards a steady-minded spiritual path. And in the fourth hand, he holds a Modak, a sweet confection made with coconuts to remind the devotee to let go of his/her ego. The breaking of a coconut is a common practice in Hindu temples and it signifies the breaking of the individual's ego (coconut shell) to expel the fluid mind (coconut water) and to expose the sweet, steady mind (coconut meat) within.

A Ganesha devotee is therefore praying to be a compassionate, inspired problem solver, while acting without desires and without ego, the two main characteristics of a Dharmic person. That is, if the devotee truly understands the symbolism of the Ganesha representation.

Lord Shiva is usually shown with a cobra wrapped around his neck, signifying his powers of mental concentration. He is shown carrying a trident with which he is said to open the third eye of wisdom by shattering the ego. The third eye of wisdom, shown in the center of his forehead, allows us to see things as they are, not as we are, by eliminating the illusions in our minds. Lord Shiva is also shown seated in a meditative pose, reminding the devotee that the mastery of desires and the dropping of our egotistic illusions can be accomplished through the practice of mindfulness and meditation.

Lord Krishna is shown playing a flute which symbolizes the state of egolessness that a devotee aspires to achieve. In such a state, where the devotee is like the flute, Lord Krishna plays through the devotee to create beautiful music. If the ego tries to play the flute, then the result is cacophony. Suffering is an indication that the ego is trying to play the flute, which should then remind the devotee to let go.

Through such symbolic depictions, Hinduism steers us to be compassionate and detached in our feelings and to act in a state of desirelessness and egolessness, to become Butterflies. And despite the numerous gods and goddesses, Hinduism is truly PanEntheism. Its core belief can be summarized as

Everything is in Consciousness (Paramatma, God); (Paramatma, God) Consciousness is in Everything.

In the Bhagavad Gita, Lord Krishna tells Arjuna, "He who sees me in all and all in me is near and dear to me," encapsulating this core belief. As such, Hindus develop a very personal relationship with God and the Hindu concept of Moksha or Heaven is right here on Earth, in this life. Moksha is attained when we drop the illusion of separateness from all Creation and achieve a state of "Ananda" or bliss in that oneness. Since all Creation is a manifestation of Consciousness and is therefore sacred, the prescription for conducting an individual's life becomes the first Mantra of SriIsopanishads:

Take just what you need from the Earth and no more, for the Earth and all its bounty does not belong to you, but to the Lord.

But that's easier said than done especially when we have such clever minds that can easily redefine wants into "needs." Therefore, a practical, non-prescriptive formula is

Act according to your Dharma, but Let Go of the fruits of your actions.

But the path of Dharma is personal. To each his own, according to his circumstances. The Hindu sages were wise enough to know that the human mind works through stories. It is our remembered stories that help us filter and process the vast inputs we receive through our senses from reality. It is our remembered stories that help us choose the path of Dharma in any circumstance. So they bombarded us with stories full of symbols hoping that some of these stories would stick. But they didn't bargain on the fact that we would forget the symbolism of the stories and cease to teach that to our children. Furthermore, they didn't count on us being ensconced in a culture where, far from letting go of the fruits of our actions, we intensely cling to them as a matter of course, hiring entire phalanxes of lawyers, if necessary, to enforce our "rights" to those fruits, especially, if we are Corporations masquerading as Persons. We have even codified our clinging to the fruits of our actions in the heart of the US Constitution via the Patent clause, signifying the importance that we attach to the clinging.

5.2 The Unknowability of Absolute Truth

In the Hindu view, the absolute Truth is unknowable. The story of the six blind men and the elephant illustrates this view. Here's John Godfrey Saxe's beautiful poem from 1872[4] relating the story:

IT was six men of Hindustan
To learning much inclined,
Who went to see the Elephant
Though all of them were blind,
That each by observation
Might satisfy his mind.

The First approached the Elephant,
And happening to fall
Against his broad and sturdy side,
At once began to bawl:
"God bless me!—but the Elephant
Is very like a wall!"

The Second, feeling of the tusk,
Cried:"Ho!—what have we here
So very round and smooth and sharp?
To me 't is mighty clear
This wonder of an Elephant
Is very like a spear!"

The Third approached the animal,
And happening to take
The squirming trunk within his hands,
 Thus boldly up and spake:
"I see," quoth he, "the Elephant
Is very like a snake!"

The Fourth reached out his eager hand,
And felt about the knee.
"What most this wondrous beast is like
Is mighty plain," quoth he;
"'T is clear enough the Elephant
Is very like a tree!"

The Fifth, who chanced to touch the ear,
Said: "E'en the blindest man
Can tell what this resembles most;
Deny the fact who can,
This marvel of an Elephant
Is very like a fan!"

The Sixth no sooner had begun
About the beast to grope,
Than, seizing on the swinging tail
That fell within his scope,
"I see," quoth he, "the Elephant
Is very like a rope!"

And so these men of Hindustan
Disputed loud and long,
Each in his own opinion
Exceeding stiff and strong,
Though each was partly in the right,
And all were in the wrong!

So, oft in theologic wars
The disputants, I ween,
Rail on in utter ignorance
Of what each other mean,

Carbon Dharma

And prate about an Elephant
Not one of them has seen!

Each one of us is like a blind man describing the elephant when it comes to our understanding of truth, reality, existence as we cannot fully grasp it due to our inherent limitations. I am always reminded of this when magicians perform their tricks demonstrating how much my senses continually misunderstand the reality around me. As such, I subscribe to the sentiments in Vaclav Havel's dictum, "Keep the company of those who seek the truth, but run from those who have found it."

5.3 The Battle in our Minds[5]

Of all the stories in the Hindu pantheon, the stories in the epic Mahabharata are perhaps the most well known and are suffused with symbolism. The Mahabharata is mainly about the battle of Kurukshetra waged by protagonists who are cousins, the Kauravas and the Pandavas, along with their vast armies. It is said that over 4 million soldiers, almost every able bodied man in India, participated in this battle over five thousand years ago and that the battle took place over 18 days at the end of which, there were only twelve survivors!

The Kauravas were 100 brothers and one sister, born to the blind king Dhritarashtra and his queen, Gandhari. After marrying Dhritarashtra, and in sympathy with his blindness, Gandhari deliberately blind-folded herself so that she could experience the same world as her husband. She consummated her marriage to Dhritarashtra and after a two year pregnancy gave birth to a big lump of flesh. Sage Vyasa then stepped in to answer Gandhari's prayers, divided up the flesh into 101 pieces, which developed into the 100 Kaurava brothers and their one sister, Dushala.

The Kauravas were basically "bad" people. They were prideful, hurtful, shameful, treacherous and used every trick in the book to gain an advantage in the stories of the Mahabharata and in the battle of Kurukshetra.

The Pandavas were five brothers born to Kunti and Madri, queens of Dhritarashtra's brother, Pandu. Since Pandu had been cursed to die if he ever had intercourse with a woman when he accidentally killed the sage Rishi Kindama and his wife during their act of intercourse, the five Pandava brothers were born through immaculate conceptions with various Gods. Yudhisthira, the eldest, was born to Kunti and Lord Dharma and was the personification of Dharma or righteousness. Bhima was born to Kunti and the Wind God, Vayu, and was the personification of Strength. Arjuna was born to Kunti and the Sky God, Indra, and was the personification of Courage. Nakula and Sahadeva were fraternal twins born to Madri and the Ashwini twins, and were the personification of Kindness and Beauty.

Pandu himself, spent the rest of his Life meditating in the forest, to repent for his act of killing Rishi Kindama and his wife.

As you can imagine, the Pandavas were the "good" side. All five of them married Draupadi, the personification of Devotion to Lord Krishna. And the true hero of the Mahabharata story is Lord Krishna, the personification of God himself. Lord Krishna was the ruler of Dwaraka and when the opposing cousins visited him to seek his help in the battle, Lord Krishna was sleeping in his palace. Duryodhana, the eldest Kaurava brother, being arrogant and considering himself Lord Krishna's equal, chose a seat at Lord Krishna's head and waited for him to wake up, while Arjuna sat at his feet. When Lord Krishna woke up, he saw the cousins and he offered them two choices. They could choose his entire army to fight on their side or they could choose Lord Krishna himself, except that Lord Krishna would not actually wield any weapons. Arjuna chose Lord Krishna who became his charioteer, while Duryodhana chose his army. Both Arjuna and Duryodhana were pleased with their respective choices.

The symbolism of the characters in the Mahabharata becomes evident when we read through its central essence, the Bhagavad Gita. Dhritharashtra, the blind king of the Kauravas represents the human mind, the subconscious mind, his wife Gandhari represents the decision-making intellect or the rational mind, while their eldest son, Duryodhana represents the human ego. That is, when a blind mind (Dhritarashtra) is married to a blind-folded intellect (Gandhari), the result is a being with a hundred and one demonic characteristics (mostly male) and with an overweening ego (Duryodhana) that considers itself an equal to God (Lord Krishna). Conversely, when a discriminating intellect (Pandu) is married to a steady mind (Kunti/Madri), the result is a just, courageous, strong, kind and beautiful being, who becomes God's devout favorite.

One fascinating character in the Mahabharata is Drona, the teacher of both the Kauravas and the Pandavas. He symbolizes the power of habits in a human being and though portrayed as a good person, Drona fights on the side of the Kauravas in the battle of Kurukshetra. That is because, psychologically, in a battle between uplifting and degrading tendencies in a human being, the power of habit usually sides with the degrading tendencies. While good habits are necessary for personal growth initially, the good qualities that result transcend habit and are inspired from deep within. Habits do not enslave a truly self-realized person, while habits can form addictions when acted upon mindlessly.

The battle of Kurukshetra is therefore, the timeless battle in our minds between our degrading tendencies and our uplifting tendencies. It is the battle between the Caterpillar and the Butterfly within us, or the Participant and the Witness with us, to use the Cosmic Fig Tree's protagonists. Once I realized this, I understood how the Bhagavad Gita could become such a source of inspiration for a truly non-violent Mahatma Gandhi, despite its setting in a gory battleground. And I also understood why Lord Krishna found it necessary to cajole Arjuna to fight this battle in the Bhagavad Gita, for our normal inclination is to give in and to let our Kauravas, the Caterpillar, run amok within us.

In the Mahabharata, the battle was eventually won by the Pandavas, the uplifting tendencies, with the aid of Lord Krishna. It is the same battle that needs to be fought today to restructure our modern industrial civilization. For, in reality, we are all the Arjunas that Lord Krishna is trying to cajole into fighting the battle of Kurukshetra, the battle in our minds, to free the Butterfly within us.

Fig. 5. *The metamorphosis of the Caterpillar into the Butterfly.*

6. The Caterpillar Culture

"As long as people believe in absurdities, they will continue to commit atrocities" - Voltaire.

The Caterpillar culture is the culture of the overweening, chest-thumping human ego. It is a culture that celebrates the Participant, the wishing child under the Cosmic Fig Tree, and works to meet its every demand regardless of the consequences to Life on Earth. It is a culture rooted in the exceptionalism of the human species that originated in the industrial West and it is now sweeping throughout the world, assimilating and devouring indigenous cultures everywhere. Says Wade Davis of the National Geographic,

> "Just as the biosphere has been severely eroded, so too is the ethnosphere -- and if anything, at a far greater rate. No biologists, for example, would dare suggest that 50 percent of all species or more have been or are on the brink of extinction because it simply is not true, and yet that -- the most apocalyptic scenario in the realm of biological diversity -- scarcely approaches what we know to be the most optimistic scenario in the realm of cultural diversity[1]."

Half the indigenous cultures of the world that existed 50 years ago are already extinct as of today, taking with them not just their unique languages but also the stories that wove the fabric of those cultures and the wisdom encapsulated in those stories.

When industrial civilization began in the 17th century, factory owners quickly discovered the advantages of standardization and specialization; standardization so that the same parts can be used for different products and specialization where workers with the most aptitude for a certain task get assigned to that task continuously. The owners' profits maximized with such specialist assignments on the assembly line. However, having to do the same thing over and over drove the workers nuts, and as a result, absenteeism was rampant in the early stages of the industrial revolution. To reduce absenteeism, the factory owners realized that they needed to keep the workers fearful and insecure about the future, and to make them dependent upon their salaries for their subsistence. They needed the workers to find themselves in an environment of scarcity. Even children were cruelly exploited to support the factory system in its early days[2].

Modern industrial civilization has progressed into a gargantuan version of this 17th century English factory, with a hierarchical, pyramidal organizational structure with individuals, corporations and institutions as components. Individuals are farmed through the educational system to find their niche within this pyramidal structure. They are "standardized" with the awarding of diplomas and degrees and their skill sets are specialized through their branches of learning. The pyramid is made up of the supply chain of resources for the products of the civilization and people position themselves all along the supply chain, feeding off the profits. Therefore, the pyramid is hierarchically layered with Nature at the very bottom, poor people most concentrated just above Nature, and the wealthy typically occupying higher layers of the structure. Wealth is extracted from Nature by the very poor and it trickles up to the top. Depending upon the level that they occupy in the hierarchy, individuals get to sample more or less of the wealth that is trickling - some would say, gushing - up to the top. But the most important component of this pyramidal structure that defines the Caterpillar culture is the modern corporation. Of the top 100 economies in the world, 51 are corporations[3]. And executives do things on behalf of their corporations that they wouldn't dream of doing to their neighbors in their private lives. For though they are accorded the same privileges as persons in the United States, modern corporations are expected to behave like quintessential Caterpillars.

6.1 Lust, Aggression and Greed

Soon after the Financial crisis of 2008 unfolded[4], I was attending a conference in Bangalore, India, where the Keynote speaker was an economic advisor to the Prime Minister of India. The speaker began by observing that greed was being touted as the main reason for the Financial crisis and that such a diagnosis was completely unhelpful. He went on to say, "Of course, greed played a part in the Financial crisis because greed is the basis of our capitalist system. It is greed that drives our corporations, not altruism." It was a frank admission of the rot in the foundations of modern industrial civilization.

My first personal experience with how large corporations behave was in 1993. We had just built our "American dream" home in Colts Neck, New Jersey, a custom home situated on over an acre of land. Our home took almost a year to build. It had a bedroom downstairs to accommodate any visiting elderly relatives and friends - it was mostly intended for our aging parents - and it had four bedrooms upstairs for us, our two children and a guest. We moved into our new home in September, just before the cold season started. I was feeling really good about our life and our prospects and a month later, I decided to buy a brand new luxury car, then drove it home and parked it in our garage.

The next morning, this brand new luxury car caught fire and burnt out a substantial portion of our home, including most of our photographs and other personal mementos. It was around 10am in the morning and our sons were playing in the bonus room above the garage while I was on the phone with a friend. Our older son, Sushil, ran out of the room and told me that there was smoke coming out of the heating vent. I thought that it must be the new heater which had just kicked in at the onset of the cold season and told him not to worry about it. Then he ran out again and insisted that there was really a lot of smoke coming out of the vent. This time, I hastily concluded my call and found that smoke was indeed billowing out of the heating vents in the bonus room. I ran down to the garage to check up on the heater that was situated inside, opened the door and saw our new car spitting flames out of its grill, while emitting loud pops and crackles. A few unopened cardboard boxes in front of the car had already caught fire and the garage was full of smoke.

I ran upstairs to the bonus room, collected our children, hit the fire emergency button on our alarm keypad and ran out of the house to wait for professional help. A policeman arrived in 5 minutes with his little fire extinguisher, but once he saw the conflagration in the garage, he dropped it, took out his camera and started taking pictures. The fire truck arrived within 10 minutes and got the fire under control within half an hour, but the damage was done.

Riding on the heels of the fire-truck came several lawyers, who were apparently professional ambulance chasers, constantly tuned into the police radio signals. They came to solicit our business, to represent us against our insurance company, to ensure that we obtain the maximum compensation for our personal tragedy.

We sent them away.

Next came the Monmouth County Fire Marshall and the Arson Unit investigator who poked through the debris in our garage and diagnosed the origin of the fire to be within the engine compartment of my new car. At which point, I called the salesman who had sold me the car to berate him. He was solicitous of our situation and was very sorry that we were yet another victim of this problem. He gave me the phone number of an engineer in the auto company, Eric, and asked me to talk to him about the issue.

I had a long conversation with Eric, engineer to engineer, and he explained exactly what happened. Our car was apparently the eighth one to ignite, but the head office was waiting for ten cases before ordering a recall. The engine compartment in our car was too cramped as it was originally designed for a six-cylinder engine, but an eight-cylinder engine was fit into it in response to market competition. As a result, the fuel line was routed too close to the heater for the windshield washer fluid and a leak in the fuel line could cause the fuel vapors to concentrate around the heater. When the vapor concentration and the ambient temperature reached appropriate levels, the mixture could ignite spontaneously, especially if the vehicle is in a closed location where there is no air circulation to disperse the fuel vapors.

As, for example, in our garage.

The next day, there were experts from both the insurance company and from the auto company sifting through the debris of our house fire looking for clues to build their respective cases. The expert for the auto company was a Professor of Physics from the University of Pennsylvania and when he found out that I had been talking to an engineer in the company offices, he became quite animated. He must have taken steps because from that point onwards, I could never reach Eric, the engineer, on the phone.

As time went on, it became quite obvious to me that the good Professor's objective was to find an alternate story for the fire. He filed a report implicating the heater in the garage. In his version of events, the heater which was in a corner of the garage, caught fire and then threw flames at the car, which was in the center of the three-car garage. The car, of course, was the main fuel for the fire as the policeman's photos captured flames bursting from the center of the garage first.

Once the professor filed his report which allowed his temporary employer plausible deniability, the insurance company sued the auto manufacturer and the resulting case dragged through the courts, requiring lots of depositions over the next four years.

Six months after the fire, we received a notice from the auto manufacturer announcing a recall on the car to put a sleeve on the fuel line to prevent fuel vapor leaks from causing a fire. The design flaw to be corrected was the proximity of the fuel line to the heater for the windshield washer fluid compartment. This was the exact same problem that Eric had described to me six months earlier and I deduced that two other cars must have caught fire since our incident. I called the telephone number on the recall notice and informed them that the vehicle in question was already burnt to its shell.

Four years after the fire, the insurance company sent us a check for 30% of our deductible as they had concluded their legal case with the auto manufacturer for 30 cents on the dollar under a no-fault settlement. Despite the recall notice, the auto manufacturer continued to deny that our house fire was caused by their product, based on the good Professor's report.

In the beginning, I was deeply disturbed at the lack of transparency of the auto manufacturer and the lack of academic integrity of the Professor. In retrospect, however, I feel that the Professor was just doing the job that he was asked to do and the corporation had no choice but to dissemble as it did because of the culture that it is ensconced in. If they had admitted to the culpability of their product for the house fire, it is likely that ambulance chasers would have come out of the woodwork to sue them for numerous other fires. In the litigious system that has evolved in the United States of America and throughout the world, ethics and morals would seem to be luxuries for corporations. But, despite the dissembling, the corporation was obviously working on the problem internally and took steps to correct it. This is similar to what corporations did when they were confronted with the stratospheric ozone hole problem due to the use of chlorofluorocarbons (CFCs) in the seventies and eighties[5]. While they were attacking the scientists who had discovered this problem, they were busy working on alternatives internally. Once they found a suitable alternative, they acknowledged the problem and changed course.

However, corporations, especially large ones, tend to put profits above morals. In the Nation's investigative article, "The Secret History of Lead[6]," author Jamie Lincoln Kitman begins,

"The next time you pull the family barge in for a fill-up, check it out: The gas pumps read 'Unleaded.' You might reasonably suppose this is because naturally occurring lead has been thoughtfully removed from the gasoline. But you would be wrong. There is no lead in gasoline unless somebody puts it there. And, a little more than seventy-five years ago, some of America's leading corporations--General Motors, Du Pont and Standard Oil of New Jersey (known nowadays as Exxon)--were that somebody. They got together and put lead, a known poison, into gasoline, for profit."

As an anti-knock agent, the compound, TetraEthyl Lead was added to gasoline instead of an equally effective compound, Ethyl Alcohol, because these companies had the patent on the lead additive while everyone knew how to make Ethyl Alcohol or "moonshine" already. This one sordid act committed by these corporations led to millions of people being exposed to lead poisoning in the affluent world. Lead contamination of the environment continues to this day with leaded gasoline still being sold in the Global South and in Eastern Europe as corporations pursue their profit motive, disregarding the public good. In America, corporations are laying off older workers and hiring younger ones at much lower pay, since experience doesn't count for much any more: knowledge and experience are freely available on the internet. Other corporations are laying off local workers and hiring guest workers from foreign countries at exploitative wages instead. The "public good" has been slowly eroded from the charter of the corporation under the guise that the public would determine whether the corporation is good or not through their purchases.

Even after the environmental movements in the sixties and seventies achieved several landmark victories with the passage of the Clean Air Act, the Clean Water Act and the Endangered Species Act in America, corporations grandfathered in escape clauses that they have clung to even today. A friend of mine, who used to work for a large chemical corporation, told me of a factory on the banks of the Raritan River in New Jersey which the corporation ostensibly has kept open until now. Except that there is just a single employee whose job it is to open the factory in the morning and close it in the evening, just so that the corporation can comply with the terms of the Clean Water Act. The factory had dumped carcinogens in the surrounding land for years and the corporation knew that it would take a lot of money to clean it up, but as long as the factory was still "open," the corporation could defer the clean up under the terms of the Act.

This is the system that we have set up. The corporation is supposed to look out for shareholder interest exclusively and these days, for large corporations, it is just the return on investment for a few large mutual and pension funds that matter the most. It is in these shareholders' interest for the corporation to externalize as much of the cost of producing goods as possible. Then the government is supposed to box in the activities of the corporation, to look out for the public good. Meanwhile, the corporation will do all it can to throw off the shackles of such government regulation, also to promote shareholder interest, such as, for example, by financing the elections of the lawmakers. When the Supreme Court of the country is also stacked in favor of the corporations, the interest of the shareholders can be well and truly addressed.

Of the three cardinal sins in most religions, Lust, Aggression and Greed, the modern corporation has to openly flaunt all of these or perish; Lust to promote its products, Aggression to try and squash its competitors and Greed to execute on its fundamental objective, to maximize its profits. An executive team that fails to commit these cardinal sins on behalf of the corporation will most likely get fired by its board.

According to the Bhagavad Gita, Lust, Aggression and Greed are the three gateways to Hell on Earth. Yet we have such powerful entities in our modern culture that are built around these very same characteristics. For we were somehow duped into believing that this is how components in natural ecosystems behave, in a crassly self-interested, hyper-competitive manner. And that somehow, the net result is benevolent to the whole ecosystem, to the whole society. This is an absurd, cherry-picked distortion of Adam Smith's original formulation of the objective function of capitalism: Enlightened Self-Interest. Our modern Caterpillar culture has forgotten the "Enlightened" part and is pursuing just the "Self-Interested" part. With narrow, unenlightened self-interest as the primary goal of commerce, the resulting economic system is what I call "Catapilism," an anagrammatic distortion of Capitalism. Adam Smith himself warned against such distortions 250 years ago when he said, "An investment is by all right-minded people to be commended, because it brings comforts and necessities to the citizenry. But, if continued indefinitely, it will lead to the endless pursuit of unnecessary things[7]."

In its section on Adam Smith's "Invisible Hand of the Market," Wikipedia has highlighted the distortion of Adam Smith's formulation in Paul Samuelson's popular Economics text book for example, showing Smith's original text and Samuelson's cherry-picked selections in bold[8]:

> **"As every individual** ... therefore, endeavors as much as he can, both to employ his capital in the support of domestic industry, and so to direct that industry that its produce maybe of the greatest value; every individual necessarily labors to render the annual revenue of the society as great as he can. He generally, indeed, **neither intends to promote the general** [Smith said "public" not general] **interest, nor knows how much he is promoting it.** By preferring the support of domestic to that of foreign industry, **he intends only his own security,** and by directing that industry in such a manner as its produce may be of the greatest value, he intends only **his own gain; and he is in this,** [as in many other cases] **led by an invisible hand to promote an end which was no part of his intention.** Nor is it always the worse for the society that it was no part of it. **By pursuing his own interest, he frequently promotes that of the society more effectually than when he really intends to promote it."**

And, voila', we have found a theoretical basis for promoting Lust, Aggression and Greed in the very foundation of modern industrial civilization. We each act in our own self-interest, maximizing security and wealth for ourselves and our own children and assume that somehow it translates into a prosperous world through the invisible hand of the market. When we attempt to construct a just society on such a basis, we are essentially attempting to prove that all our religious texts are just plain wrong. It seems to me that as long as we believe in such absurdities, we will continue to commit not just atrocities, but also hara-kiri.

6.2 War on Nature

Yet another absurdity that forms the basis of modern industrial culture is our assumption that we humans can dominate and bend

Nature to our will. In an eponymous book, "Domination of Nature[9]," author William Leiss traces the destructive impact of modern civilization on Nature to pervasive attitudes originating with Sir Francis Bacon in the 17th century. Countering the defeatism in 17th century society towards natural disasters in the form of the bubonic plague, childhood diseases, etc., Bacon proposed that the mechanical inventions of the industrial revolution be employed to "conquer and subdue Nature, to shake her to her foundations." Thus Bacon fired the first salvo in humanity's "War on Nature," the longest running, continuous war to date, far surpassing the "War on Poverty," the "War on Drugs" or the "War on Terror" in terms of ferocity, sheer breadth of social complicity and utter absurdity. The modern Caterpillar culture is a realization of Sir Francis Bacon's metaphoric view of Nature in shackles being enslaved in service of humans. In his book, Novum Organum[10], he wrote, "I am come in very truth leading to you Nature with all her children to bind her to your service and make her your slave." But if Nature is the "slave" of humans, then extinction and climate change would constitute the "slave rebellion."

It was perhaps understandable for 17th century Europeans to embark on such a "War on Nature" as this was the age of superstitions, leach doctors and witch trials in Western society. That this war has continued on for more than four centuries to become part of the dominant culture today speaks to the thorough brainwashing that society undertakes on children as it propagates its traditions. The Baconian "War on Nature" is not a war on Earthquakes, Volcanic Eruptions, Hurricanes, Tornadoes, Floods, Droughts and other natural phenomena, but a war on the weaponless fellow beings who inhabit our planet. Indeed, the modern fishing industry employs all the tools of modern warfare: satellite imagery, GPS signals, remote monitors, to corral the prized fishes of the ocean. The captains of modern fishing fleets are more Captain Kirks than Captain Ahabs, to use Jonathan Safran-Foer's analogy from his book, "Eating Animals[11]." At a more gut-wrenching level, it is traditional in Canada to club hundreds of thousands of baby seals to death on the beaches of Newfoundland each year[12] and it is traditional in the Faroe Islands of Denmark for teenagers to engage in the wanton slaughter of Calderon dolphins as a rite of passage into adulthood[13].

While the Canadians and the Danes get the brunt of bad publicity for their overt display of cruelty in the name of tradition, theirs is but a very tiny part of the holocaust that humans routinely commit upon other species. Almost 60 billion land animals are raised and killed each year to satisfy human appetites, with 10 billion of them in the US alone. In addition, some 90 billion sea creatures are killed each year for food[14]. Most of the land animals we consume are raised in factory farms amid squalor, in cramped quarters and they are given a steady diet of antibiotics to overcome the illnesses that result from such conditions. Humans use animals for pets, food, clothing, entertainment and research and in each of these realms, millions and billions of animals suffer. Even the pet industry is full of suffering. Millions of unwanted pets are housed in cages in dank animal shelters before being euthanized and "rendered," that is, ground up, cooked and processed, along with all the other "recycled" animal body parts, into livestock feed. In the US, twice as many animals are euthanized than adopted from shelters. Even the venerated cows in India are forced to walk to their demise in leather tanneries through the judicious sprinkling of spicy chili powder in their eyes whenever they stall. The People for the Ethical Treatment of Animals (PETA) has documented the horrific conditions that circus and zoo animals endure in the entertainment industry[15]. In the pharmaceutical industry, the Food and Drug Administration (FDA) insists on animal testing before approving drugs and treatment procedures even though such testing is cruel and has become increasingly archaic as there are far better computer-aided techniques available. The documentary, Earthlings[16], that is freely viewable over the internet goes through in excruciating detail the misery and suffering that animals undergo to support the human enterprise that is modern industrial civilization.

Unfortunately for the tiger (rhino), its penis (horn) is said to have tremendous invigorating potential in Traditional Chinese Medicine[17]. As a result, poachers hunt down male tigers (rhinos) in order to grind their penises (horns) into powder for the benefit of the affluent, but impotent humans. The bile from the gall bladder of live bears is also used in Traditional Chinese Medicine and there are bear bile factories in China where the bears are kept alive in iron contraptions called crush cages while the bile is drained from them. These crush cages are so small that the bears are unable to move their paws and injure themselves while trying to overcome the pain of the process. Recently, there was a report that a mother bear freed herself from her crush cage upon hearing her cub cry out in fear when the workers were about to insert the tube into the cub's gall bladder[18]. The mother bear then chased away the workers, suffocated her cub to death before banging her head against the wall and committing suicide. She rid herself and her loved one of a life of torture.

War is hell. Humans are winning the war and to the victors go the spoils, don't they? Around half of the photosynthetic output of the planet now goes to support humans with the other millions of species having to make do with the remainder[19]. The proportion of the Earth's photosynthetic output going to support humans is increasing every year as more and more humans are demanding the luxuries that the affluent take for granted today. Some people even justify this on the grounds that every species is always trying to outcompete others and grab as many resources from Nature as it can and that humans are no exception. That Nature works on the principle of the survival of the fittest and humans are clearly the fittest of all the species.

The trouble with this argument is that "survival of the fittest" is a gross distortion of the principle of evolution by natural selection that governs ecosystems behavior[20]. If the fittest species in the forest, say the tiger, decides to kill everything in sight because it could, then that forest would die and with it, the tiger. The tiger doesn't do that which is why you can see a herd of deer grazing within sight of a tiger that has eaten its fill for the day. In contrast, when one species outgrows and overwhelms all other species in an ecosystem, that is actually a sign of an impending collapse of that system. It is symbiosis, the give and take of Life, that governs the behavior of species in a healthy ecosystem, not "survival of the fittest."

Oftentimes, we convince ourselves that we are substituting one species for another and are therefore, a benign influence on ecosystems. During the "Taming of the Wild West[21]" in 19th century America, countless natural treasures of the continent were destroyed in the name of civilization. The buffalo hunt of the West was intended not just to subjugate the indigenous cultures that depended on the buffalo but also to free the land of predators and make it amenable for the cattle herds of the European settlers. Passengers used to shoot buffaloes from passing trains just for fun, and together with the professional hunters such as Buffalo Bill Cody, they wiped out an estimated 25 million buffaloes in about two decades. Nowadays, more than half the land area of America is used for livestock production. Cattle herds were supposedly a ruminant substitute for what the buffalo herds used to do. However, while the buffalo herds supported entire ecosystems of predators and prey in the original Great Plains, the cattle herds only support humans. For any wild animal that has the temerity to consume cattle is mercilessly hunted down by humans and murdered.

From the outset, Science had a major role to play in this War on Nature. Science developed the tools for conducting the War on vaster and vaster scales. The Amazon is such a tremendous store of biodiversity that it is almost impossible to deforest portions of it without it growing back. There's a saying among the native tribes of the Amazon that it has more eyes than leaves, which is amazing considering that there are 600 billion large trees in the Amazon. Yet, through the use of modern technology, we pour herbicides on the deforested Amazonian land to kill off everything that would grow and then plant genetically modified soy that has been engineered to be resistant to that herbicide. The GMO soy grown in the deforested Amazon is then used to feed livestock for the growing appetites of human beings.

In the ocean, GPS and other satellite technology are used to track and trap rare fish stocks even as the stocks decline and prices rise astronomically. On purely cultural considerations, in March of

2010, the nations of the world voted to condemn the Bluefin Tuna to extinction at the Doha round of the Convention on International Trade in Endangered Species (CITES), so as not to offend the sensibilities of Japanese sushi consumers[22]. In January of 2011, a single bluefin tuna fetched a record price of $396,000 in a Tokyo auction[23]. The appetite for tuna and other top predatory fish hasn't declined among the rich, even as the concentration of mercury, carcinogens and radioactive waste inexorably increase, year by year, in the bodies of these creatures.

It is the "scientific management[24]" of forests that led to the decimation of native species throughout India during the British rule in the 19th century, with the wholesale substitution of Pine trees for Cedar trees in the North and Teak trees for JackFruit trees in the South. The objective of the "scientific management of forests" was to maximize the utility of the forests for human needs at the expense of native flora and fauna. Unfortunately, the substitution of pine trees for cedar trees in the Himalayas wiped out the incredible diversity of plants and herbs that the cedar forests supported since pine cones cover up the ground and prevent other plants from sprouting. Likewise, the teak plantations are useless for the elephants who eat jackfruits, spread the jackfruit seeds and keep the forests of the Western Ghats of South India flourishing with their presence. While indigenous cultures had subsisted in the forests for centuries using the concept of sacred groves in the North and "DevaKadu" or "God's forest" in the South to preserve the biodiversity of the forest, their sustainable means of livelihood were wiped out in this sweep of scientific management. Science is a wonderful tool when wielded with compassion, but it is a lethal tool in the hands of Caterpillars.

While the War on Nature was conducted through Science, it was justified through religion. Certain religious interpretations deified Homo Sapiens as the only species on the planet that was made in the image of God and that possessed souls while all other species were considered soulless automatons who were put on earth for human use and enjoyment. When I was talking about veganism to a friend of mine recently, he blurted out, "But animals have no souls!" meaning that animals are simply like vegetables without thoughts or feelings or capacity for suffering. For someone who has lived with dogs and cats all my life, it was astounding to hear such a viewpoint from an educated person. In this world view, it isn't just animals that are accorded this denigrated status, but all the trees, plants, birds, fishes, insects and even entire forests as well. This is why it is considered acceptable in our civilized society to blow off the tops of mountains in order to scoop out the mineral ores underneath and then cover up the resulting hole with sod to supposedly restore the destroyed greenery.

But as we have become very effective at killing off our fellow species, we are only now beginning to realize that we are essentially killing ourselves. For Sir Francis Bacon's War on Nature is really a War on humans. Shaking Nature to her foundations results in the rattling of teeth in our own skulls.

It is absurd.

6.3 Winner Takes All

The Caterpillar culture is built on the use of "winner takes all" style competitions as the engine for progress. This is so fundamental to the culture that a patent and copyright clause "to promote the Progress of Science and useful Arts, by securing for limited Times to Authors and Inventors the exclusive Right to their respective Writings and Discoveries," was written into Article 1 of the United States constitution[25]. It doesn't matter if the inventor is standing on the shoulders of countless giants and takes just the last tiny step to develop the invention; he still gets to keep all of the profits from that invention once the patent is granted. He won the horse race! Of course, it is only for a limited period, currently 17 years. These days, it is probably the inventor's large corporate employer who gets to enjoy those profits for 17 years. But 17 years is now a very long time as product cycles get shorter and shorter.

"Winner takes all" systems create monopolies. Monopolies wield power and power begets monopolies in a self-reinforcing cycle. At the beginning of the 20th century, it was John D. Rockefeller and his Standard Oil Corporation, now ExxonMobil, that wielded such power. It required a dynamic President like Teddy Roosevelt to bust up that monopoly and put in place anti-trust laws to prevent future monopolies[26]. But increasingly lax enforcement of anti-trust laws in both Republican and Democratic US administrations has been a godsend for the new near-monopolies of the late 20th century and early 21st century.

"Winner takes all" is so ingrained in our modern industrial culture that we think nothing of it even when it results in the flagrant display of insensitivity and even cruelty. There is a very popular Indian television program along the lines of "American Idol," called "Little Champs[27]," where little children sing their hearts out to a worldwide audience. Advertisers clamor to sell their products on this program and the angelic children do a wonderful job captivating the audience. However each one of these programs invariably ends with one child sobbing in tears as he or she is eliminated from the competition. For, according to the judges of the competition and the worldwide audience, that sobbing child was just a tad bit less talented than all the other children on stage. After making every child, except one, cry during the entire season, the winning child is crowned and that child can then be found selling products for advertisers in the following years of the program. And we call this entertainment, while we justify the public humiliation of these beautiful children as a character building activity.

Due to the "winner takes all" paradigm, in almost every profession, we now find "superstars" earning untold riches, while the majority of the practitioners languish in the basement fighting over the crumbs. The new communications technologies, the internet, cable and satellite TV, wireless and WiFi technologies, have perversely increased the audiences and hence the wealth accruing to superstars, the best of the best sports heros, movie and television stars, celebrity chefs, fashion models and music icons. These superstars are then used to hawk products to the masses and redirect wealth to the executives and CEOs of major corporations who can afford to employ them. The result is a widening of the wealth gap between the rich and the poor to the point where the rich have become the dominant drivers of demand in many large economies around the world. With these widening gaps, the global economy has now truly become a non-sustainable pyramid scheme. According to Wikipedia[28],

> "A pyramid scheme is a non-sustainable business model that involves promising participants payment, services or ideals, primarily for enrolling other people into the scheme or training them to take part, rather than supplying any real investment or sale of products or services to the public. Pyramid schemes are a form of fraud."

Pyramid schemes are illegal in many countries including Australia, Canada, the United Kingdom, and the United States. Yet a strong case can be made that the entire global economy, as it exists today, is actually the mother of all such pyramid schemes[29].

The global economy is based on extracting minerals, timber, and living organisms including fertile soil from Nature and for the most part, returning carcinogens, pollution and radioactive waste back to Nature. That is, the main depositor in the global economic Pyramid scheme, Nature, is not getting any interest on the deposits and, in fact, is receiving injuries instead. The people who are enjoying the most benefit from the global economy are high up in the pyramid and thus, far removed from the depredations being done on Nature, the depositor. In his book, "The Ascent of Humanity[30]," Charles Eisenstein correctly traces the coming convergence of environmental crises to the human ego, the sense of separateness from Nature and from reality. The long supply chains of modern industrial civilization necessarily lead to such separateness as those individuals and entities that are high up in the pyramid don't have much sense of all the transformations that Nature had to undergo to create the products that they consume. I doubt that the people who are busy waving a finger over their smart phones while walking down a city street are fully aware of all the minerals that had to be extracted and processed to create the gadget that they are interacting with. Indeed, they are not even fully connected with the person that they are interacting with over the electromagnetic spectrum, even as they are not fully connected with the people surrounding them in the city street.

For technology can promote separateness while providing the illusion of connectedness. While we each have lots of Facebook friends, we have become too busy to sit and eat meals together at home.

However, I believe that this separation is a consequence of the various absurdities that we have bought into as an industrial culture and not a fundamental feature that evolves with the rise of intelligence. There are cultures deep in the Amazon and in the forests of India who are living sustainably in harmony with Nature and these cultures don't find the compelling need to separate themselves from Nature, despite their intelligence.

While the vast majority of people who are making the most demands on Nature are separated from the consequences of their demands, the people who are actually doing the dirty work of extracting the deposits from Nature, the slaughterhouse workers, the farm hands, the miners, the shepherds and the lumberjacks, tend to be poor laborers who are forced to do such work out of necessity, not choice. They are driven to this necessity as the global economy destroys the environment around them, reducing the natural bounty that their ancestors enjoyed into their present day scarcity. The Caterpillar culture depends on this scarcity and the ever increasing population of the poor - and the limitless bounty of Nature - to fuel the expansion of the base of the pyramid and consequently, the wealth that flows to the top. Unfortunately for the Caterpillar culture, the bounty of Nature is turning out to be not so limitless after all.

To ensure that there is sufficient demand for products and services to fuel the growth of the global economic pyramid scheme, the Caterpillar culture depends on another great, big Lie, a whopping absurdity. That absurdity is the 20th century Freudian assertion that human happiness can be attained through the fulfillment of desires, a lie that the Rig Veda dispelled in the Cosmic Fig Tree story and which the great Buddha shouted himself hoarse about 2500 years ago. Clearly Sigmund Freud and his lineage of psychologists were oblivious to the contents of the Rig Veda or to the Buddha's teachings, but the Freudian theories were cleverly used by corporations to create the consumer societies of the 20th century. At present, consumption accounts for 70% of all the economic activity in the US, which is why President George W. Bush and many Western leaders advised their countrymen to go shopping as a patriotic duty[31], right after the Al Qaeda attacks of September 11, 2011.

But for a brief period, in 1968, this consumer culture appeared to be in its last throes.

In March of 1968, Senator Robert F. Kennedy put it eloquently while addressing students at the University of Kansas during his bid for the Presidential nomination of the Democratic Party[32],

> "Too much and too long, we seem to have surrendered community excellence and community values in the mere accumulation of material things. Our gross national product ... if we should judge America by that - counts air pollution and cigarette advertising, and ambulances to clear our highways of carnage. It counts special locks for our doors and the jails for those who break them. It counts the destruction of our redwoods and the loss of our natural wonder in chaotic sprawl. It counts napalm and the cost of a nuclear warhead, and armored cars for police who fight riots in our streets. It counts Whitman's rifle and Speck's knife, and the television programs which glorify violence in order to sell toys to our children. Yet the gross national product does not allow for the health of our children, the quality of their education, or the joy of their play. It does not include the beauty of our poetry or the strength of our marriages; the intelligence of our public debate or the integrity of our public officials. It measures neither our wit nor our courage; neither our wisdom nor our learning; neither our compassion nor our devotion to our country; it measures everything, in short, except that which makes life worthwhile. And it tells us everything about America except why we are proud that we are Americans."

Carbon Dharma

But the mass movements of the sixties that could have blossomed into a full fledged reversal of the Caterpillar culture depended upon the charisma of leaders such as Senator Robert F. Kennedy and the Reverend Martin Luther King, Jr., to sustain. Unfortunately, soon after the Kennedy speech in Kansas, in April of 1968, the Reverend Martin Luther King, Jr., the icon of the civil rights movement in America, was assassinated[33].

In May of 1968[34], the largest ever wildcat general strike in the world brought the industrial economy of France to a virtual standstill. The strike involving 11 million workers over two weeks, began with the staging of student protests, and almost caused the collapse of the government of Charles De Gaulle. At its heart, it was a revolt against modern consumer and technical society. Says Wikipedia,

> "Many saw the events as an opportunity to shake up the "old society" and traditional morality, focusing especially on the education system and employment. It began as a long series of student strikes that broke out at a number of universities and lycées in Paris, following confrontations with university administrators and the police. The de Gaulle administration's attempts to quell those strikes by police action only inflamed the situation further... The protests reached such a point that government leaders feared civil war or revolution. De Gaulle fled to a French military base in Germany, where he created a military operations headquarters to deal with the unrest, dissolved the National Assembly, and called for new parliamentary elections for 23 June 1968. Violence evaporated almost as quickly as it arose. Workers went back to their jobs..."

Shortly after midnight on June 5, 1968, Senator Robert F. Kennedy was assassinated[35]. On June 23, 1968, the Gaullist party emerged stronger than before in the national elections in France[36].

And the Caterpillar culture got its second wind.

In 1971, the eminent barrister Lewis Powell, who months later became a justice on the US Supreme Court, wrote a memo to the US Chamber of Commerce detailing the steps that corporations and especially the Chamber, need to take to combat the environmental movement and the corporate bashing populism of consumer advocates such as Ralph Nader[37].

And the rest is history. The Caterpillar culture continued on, stronger than ever.

The Caterpillar culture assumes that Homo Sapiens is the divinely selected, pre-ordained winner in the ultimate "winner takes all" competition, the one between species. And as the winner, Homo Sapiens is entitled to take all of the resources of Nature, except in designated Nature preserves set aside by governments. And even those Nature preserves come under enormous pressure when a desirable mineral is discovered under them and politicians usually relent and let the business interests exploit those resources. It is a rare political decision when the economy does not trump the environment, even in countries where the environment is already perilously close to collapse.

Contrary to common belief, "winner takes all" style competitions are not necessary for achieving progress. In fact, most of the significant inventions of the 20th century were made by scientists working in Bell Laboratories[38] and other research institutions throughout the world. The inventors of the transistor, the laser, the cell phone and the internet were not motivated by a desire to make a lot of money. They had a much better motivator: curiosity and the fun of discovery. Besides the "winner takes all" approach becomes a formula for ensuring that both the winner and everyone else eventually take nothing. For, when applied ad infinitum, it diminishes the bounty of the planet and reduces it down to zero. The base of the pyramid of the global economy has already destroyed half the forests on land and much of the marine life in the ocean, mostly within the past fifty years, and it is still attempting to grow its footprint exponentially within the same destructive paradigm. It is inevitable that this pyramid scheme will face the exact same ending as every other pyramid scheme - it will dissolve. It is not a question of whether it dissolves, but when.

Nevertheless, at the moment, the individuals and entities that are perched high up in the pyramid are reluctant to voluntarily let it dissolve and left to their devices, will probably keep it going until it collapses. It is mainly the large corporations that have cornered the riches flowing up the pyramid that are strongly motivated to maintain the status quo. Our democratic societies are attempting to walk that fine line between Communism where the government controls corporations and Fascism where the corporations control the government and this is not always easy. For individuals high up in the pyramid, along with the separation from Nature comes the fear of not knowing what the reconnected state would look like. And fear leads to paralysis.

Perhaps it is this fear-driven paralysis that led President George H. W. Bush to proclaim to the world in 1992 that "The American Way of Life is Non-Negotiable." President Bush's stance was quite popular in America and the US Senate even voted 95-0 to reject the Kyoto protocol to mitigate climate change[39], in a 20th century equivalent of a "Let them eat cake" moment. But the corporations that control the power structures in the halls of Washington are exporting this American Way of Life to the whole globe in an attempt to increase the wealth flowing up the pyramid. It is unlikely that they will stop until Nature gets exhausted and turns the tap off.

But what if we the people, stop believing that it is in our interest to prop up those at the top of the pyramid? Then, we can voluntarily dissolve the Caterpillar culture.

But to dissolve the Caterpillar culture, we must first stop believing in the absurdities underlying it.

7. The Butterfly Culture

"You never change things by fighting the existing reality. To change something, build a new model that makes the existing model obsolete" - Buckminster Fuller.

Since around 1960, the Search for Extra Terrestrial Intelligence (SETI) project has been scanning the electromagnetic spectrum using radio telescopes, looking for signatures of intelligent alien Life from elsewhere in the cosmos[1]. The SETI@home project, which started in 1999[2], uses a network of 5 million independent home computers to analyze the data in what could be construed as the largest supercomputer ever assembled. Private computer owners just have to download some software to donate the idle cycles on their computer to be used for the search. But so far, despite all that number crunching, not a single peep has been heard that indicates the presence of intelligent alien Life in the cosmos.

However, scientists have found plenty of planets. The first extrasolar planet or exoplanet, a planet outside the solar system, was discovered in 1992[3]. Since then, over 500 such planets have been detected. Based on the data gathered so far, scientists estimate that there are at least 50 billion planets in the Milky Way galaxy alone! And there are anywhere from 100 billion to 500 billion galaxies in the observable universe[4].

That's such a mind boggling number of planets, without even a single pulse of intelligent alien Life that we could detect in nearly thirty years of searching. No doubt, the planets are full of materials that we can mine for our gadgets and products, if we can transport them. There's even a whole planet made entirely of diamonds, orbiting a pulsar in the constellation Serpens, about 4000 light years away[5]. This planet is five times larger than the Earth in diameter and 300 times heavier. It could even be just one huge diamond that would make the Kohinoor diamond seem like an inconsequential speck.

Even science, that most rigorous of all knowledge disciplines, is now telling us that intelligent Life is much rarer than diamonds in the universe.

The Butterfly culture is the culture of Life. This goes beyond sustainability. The UN Brundtland commission defined sustainable development as "development that meets the needs of the present without compromising the ability of future generations to meet their own needs[6]." This is why sustainability is commonly understood to be about maintaining the status quo on planet Earth, about living without making more of a mess than we have already made. But the Butterfly culture is about cleaning up the mess, about actively undoing the damage that has been done to the biosphere of the earth, while recognizing that it may not be possible to maintain the status quo of a damaged planet. It is about "Gardening back the Biosphere[7]," to use the words of the Indian environmentalist, Suprabha Seshan. It is about "dancing with Nature" instead of "enslaving Nature."

In Nature, Life and the environment evolved together and were perfect for each other before we came along with our dynamite, our bulldozers, our industrial processes and our chemicals. It was Life that sequestered the arsenic, the lead and the mercury in the coal seams that we are digging up today and releasing into the biosphere. It is Life that will have to trap all the poisons that we have dispersed throughout the environment and re-sequester them back into the bowels of the Earth. Trees do precisely that, storing the pollutants in their trunks and transpiring the filtered ground water into the atmosphere. These trees will eventually die and become the coal seams of the future with these poisons stored away, once again, deep underground. Therefore, it is trees and especially forests that we need to regenerate in order to repair the damage that we ignorantly did during the Caterpillar stage of our industrial revolution.

7.1 Regenerating Life

In an inspiring lecture at TEDx Amazonia in Nov. 2010[8], Prof. Antonio Donato Nobre was explaining the "Biotic Moisture Pump" that keeps the Amazon rainforest so lush. If there is a forest on land and a sea nearby, the moisture evaporating from the sea is sucked up by the air above the forest, condensed due to the microbiota that the forest emits, resulting in rain over the forest. The forest then purifies the rain water from the soil and returns it back to the sky through transpiration completing the pump action. Conversely, if there is a desert on land and a nearby sea, the air above the desert is sucked up by the moisture evaporating from the sea thereby keeping the desert trapped in that condition. This is why, after a certain level of deforestation, scientists predict that the Amazon region will be driven to desert conditions as the Biotic moisture pump shuts down and reverses course, perhaps making the region unsuitable even for GMO soy cultivation. On the other hand, if we can regenerate forests even in desert regions, we can establish the biotic moisture pump working in other parts of the world and not only undo the damage done to the biosphere but reverse climate change as well, since forests sequester carbon from the atmosphere.

Later, in that same talk, Prof. Nobre speaks of an encounter with a member of the Ianomami tribe, Davi Copenaua, who said in essence,

> "Doesn't the white man know that if he destroys the forest, the rain will end? And, if the rain ends, there will be no drinking water or food?"

Prof. Nobre continues,

> "I heard this and my eyes welled up in tears. I've been studying this for 20 years, with a super computer, with tens and thousands of colleagues to reach a conclusion that he already knows. A critical point is that the Ianomami have never deforested. How could he know the rain would end? That bugged me and I was befuddled. How could he know that? Some months later, I met him at another event and said,
>
> 'Davi, how did you know that by destroying the forest, the rain ends?'

He replied, 'The Forest Spirit told us.'

For me, this was a game changer. I said, 'Gosh, why am I doing
all this science to reach a conclusion that he already knows?'
Then something absolutely critical hit me. It is that,

Seeing is believing. Out of sight, out of heart. This is a need
that we have in our culture, that we need to see things. We live
in ignorance. Then it hit me, what if we turn the Hubble upside
down, to see down here rather than the ends of the universe?
We now have a practical reality, that we are trampling on this
wonderful cosmos that shelters and houses us, in ignorance...
Then, let's turn the Hubble around to look at the Earth, to look
at the Amazon. Let's dive in and reach out to the reality that we
live in daily and look at it, since that's what we need. Davi
Copenaua doesn't need this. He already has something that I
think I missed. I was educated by television and I missed this,
an ancestral record, a valuation of that which I don't know,
which I haven't seen. Davi is no 'doubting Thomas.' He
believes with veneration and reverence in that which his
ancestors and the spirits taught him. As we can't, let's look into
the forest with our telescopes and understand what sustains us."

Prof. Nobre captured in essence what the Butterfly culture is all
about. It is not about everyone in the industrial world returning to a
hunter-gatherer, back-to-Nature lifestyle, to become once again the
Ianomami that we sprung from. If 7 billion humans all begin to
lead hunter-gatherer lifestyles, we would most definitely wind up
hunting each other as in the multi-player video games that are
popular over the internet. That would be catastrophic. The planet
simply cannot support 7 billion predatory hunter-gatherers,
especially in its present damaged state.

Instead, the Butterfly culture is about using the tools that we have
developed towards a larger common purpose, the purpose of
healing the wounds that we have inflicted on the biosphere during
the Caterpillar stage of our global, industrial civilization. It is about
using those tools to look inwards instead of peering outwards. It is
also about realizing that the scientist is just another blind man

describing the Elephant that is Truth and not the only man who can see the Elephant. Besides, due to intense specialization, the typical scientist is only peering at the very minutest part of the Elephant these days, a toe nail here, an eyelash there. In our scientific and engineering community, we need to develop a little humility about that, just as Prof. Nobre exemplified.

I was on an airplane returning back to San Francisco from Chicago and seated next to me were a mother and a daughter. The daughter was on a semester break from college and she was talking to her mother about her college life and the courses that she was planning to take, etc. I was busy working on my laptop until I heard the daughter say to her mother in a dejected tone, "Mom, we all know that the planet is screwed, that when I grow up to be your age, I'll be living in a Mad Max world." Then I spoke up and said to the daughter to keep the faith that she won't be reduced to such a violent lifestyle if, as a Miglet, she takes steps today. But it was difficult for me to convey the gist of this book to them in just a few minutes with the result that the mother became very defensive about the American way of Life and about consumerism in general. She retorted,

> "Well, I notice that you are flying in a plane and you are typing on a laptop. Aren't you being hypocritical to use the products of our culture while criticizing it?"

I replied,

> "Actually I am part of this culture as I designed the communications system that connects this laptop to the internet. But, nowadays, I use these products for my purpose, which is to help regenerate Life. And for that purpose, I think that we must use every tool that our technologies have developed and to also develop new tools as we need them."

We must not simply discard the knowledge that we have accumulated and the technology that we have developed, but we must learn to use them for the right purpose, as we have plenty of clean-up work to do, to regenerate Life back in most environments.

7.2 Conscious Consumption

Most of us in the modern industrial world live in cities and our cities are growing. There are more people living in urban areas today than there are in rural areas. As someone living in an urban area and as a conscious consumer, I consider the most important thing that I can do to promote a Butterfly culture is to consume an Organic, plant-based Vegan diet and I have been quite vocal about my reasons for this choice. With an Organic Vegan diet, I am eating low down in the food chain, minimizing my land use footprint while also minimizing the radioactive hot particles, the pesticides and other carcinogens that I'm ingesting since all these pollutants concentrate by an order of magnitude at every step up in the food chain.

According to Prof. Danny Harvey of the University of Toronto, the total energy required for a plant-based vegan diet is approximately one-fourth the total energy required for the average world diet today and one-seventh the energy required for an affluent diet[9]. Currently, the average world diet includes about a billion people who are literally starving, bringing down the average. Including the energy input required for growing the food, an affluent family of four would save 4 times the energy switching from an affluent diet to a vegan diet than by switching from their automobiles to bicycles. Now can you imagine returning three-fourths of the land that we are currently using for livestock production back to Nature to regenerate forests? That would be a true world-changer!

Further by consuming an Organic Vegan diet, I am also assured that with every mouthful, I'm actually helping the soil to regenerate, to sequester more carbon. In addition, such a diet is not only good for the other species on the planet, but it has been good for my health as well. I consider it the least that I can do as an urban dweller to promote Life and the Butterfly culture.

As Prof. James McWilliams has pointed out[10], our diet is an intensely political choice. What all of you eat affects me, my children and especially, my grandchild, Kimaya. Conversely, what I eat affects all of you and indeed, all of Life. I have no business eating things that cause you and especially your children and grandchildren needless suffering in the long run. Organic Veganism is one of the most powerful political responses that we can make to the depredations of the industrial food system. It strikes at the very foundation of the Caterpillar culture. Besides it is hard for me to promote the Butterfly culture while eating eggs, for example, and thereby causing two day old, male, baby chicks to be thrown into a meat grinder. In the Butterfly culture, the culture of Life, babies are precious. It is also hard for me to promote the Butterfly culture while eating conventional foods where pesticides, i.e., poisons, were used to kill insects in vast numbers, thereby killing birds that feed on these insects, while those poisons then run off and pollute our waterways, killing the fishes.

Though Prof. Danny Harvey calculated the total energy required for a lacto-vegetarian diet to be just 50% more than the total energy required for a plant-based vegan diet, his calculation assumed that there are meat eaters around who would consume the spent dairy cows and their male offspring to reduce the impact of the dairy consumer. If the diary cows were allowed to live out their lives, then the total energy required for a lacto-vegetarian diet would be comparable and even higher than that of the average world diet today.

As Emerson famously said[11], "You have just dined, and however scrupulously the slaughterhouse is concealed in the graceful distance of miles, there is complicity." Extend that beyond the slaughterhouse to the baby chicks, the calves and their mothers, the insects and the birds that feed on those insects and the fish that die from the pesticide runoff and the fish that are caught to feed our appetites, to get an idea of the true complicity in the current world diet.

But such advocacy invites some very vocal detractors. Dr. Rajendra Pachauri, the head of the UN IPCC was also lambasted in the press for similar reasons, for promoting a vegetarian diet. A typical rejoinder from my detractors goes like this:

"So... is it your wish that family/small business farms and corporate farms to close down - no more cattle, pork, chicken, lamb, fish hatcheries, etc.? And no more agricultural field lots for said animal farm lots? And resultant loss of jobs that support this industry? What do you suppose the result will be to the economy? Are you willing to stand up and take the blame for crashing the economy?"

This very same argument, that change will take away jobs and crash the economy, is used to fight the conversion of our energy infrastructure from coal, oil, gas and nuclear to solar and wind as well. In a John Klossner cartoon[12] that won the Scientific Integrity Editorial Cartoon Contest for the Union of Concerned Scientists, the representative of a corporation is shown with a briefcase visiting with a family, all of whom are wearing gas masks, including the dog. In the background is a factory spewing black smoke into the atmosphere, presumably the reason for all those gas masks. And the corporate type says, "Think of how many healthcare-related jobs we're creating."

Which just about sums up the gist of this particular argument. In the documentary, "Forks Over Knives[13]," Prof. Colin Campbell of Cornell University says that if everyone in America switched to a whole-foods, plant-based vegan diet, the health care costs in America would reduce by 70-80% as incidences of chronic diseases such as diabetes, heart disease and cancer diminish. But, since health care is a $2.5 Trillion a year industry[14] in America today, think of how many healthcare-related jobs that would eliminate!

Anthony DeMello once said, "We don't really fear the unknown; we fear the loss of the known[15]." We know that there are jobs available raising, milking, slaughtering, refrigerating, packaging and transporting all the cows, pigs, sheep, chicken, lamb and fish and for growing and trucking food and antibiotics to them. In fact, almost half the food we grow is to feed our animals, not humans. There are jobs in the health care sector to care for all the cancer and heart patients that result from the consumption of animal foods. And we are scared of losing all these known jobs. We just cannot imagine how people can be gainfully employed regenerating Life while sequestering carbon, when so many are employed today killing Life on such a vast scale while emitting methane and other greenhouse gases and while destroying millions of acres of forests each year to accommodate the ever growing demand for these animal foods. And in the grip of that fear, it never really occurs to us that killing Life on a vast scale while emitting methane and other greenhouse gases and destroying millions of acres of forests each year is actually not such a smart thing to do, to begin with. That is, if we believe in the purpose of the Butterfly and we want Life to flourish on this planet.

Another typical retort is that I'm being extremist for switching to an Organic Vegan lifestyle. Why not choose a moderate, Buddhist, middle way such as Meatless Mondays[16] instead? Why not switch to "sustainable meats" such as grass fed beef instead? Even PETA presentations warn us, Organic Vegans, to consider the fact that we're not entirely independent of animals since the manure needed for growing our organic produce comes from livestock. Some of my colleagues at the Climate Project argue that if there is a substantial price on carbon, then the consumption of animal products and fossil fuels will automatically moderate. Therefore, we should all be focusing on promoting legislative action for putting a price on carbon without getting distracted with human behavioral issues.

I admit that I tend to be binary about human behavioral issues. Once I'm convinced that something is wrong, I drop it. If there are absurdities underlying our culture, then using price signals to moderate behavior leaves these underlying absurdities intact. Any changes that occur in society through such means will be due to external manipulation and therefore, not long lasting. I believe that it is far better to expose these absurdities and inspire changes from deep within each individual. For instance, putting a price on carbon will make no difference to the lifestyles of the rich, while pricing meat and gasoline beyond the means of the poor. This can be partly mitigated by returning the collected carbon tax back to the poor as a dividend, but such a proposal will run into rabid opposition in many countries where this will be construed as wealth sharing and therefore, a devious, socialist plot. Without this dividend, by putting a hefty price on carbon, we would be forcing the poor and the middle class to stop eating meat and burning fossil fuels, thereby ensuring that there is plenty to feed the appetites of the rich. Which has all the makings of an elitist plot. And we dither between a "socialist plot" and an "elitist plot," debating on the precise parameters of the tax and the dividend approach to make it equally unpalatable to the advocates of both viewpoints. And that's what external manipulation does, when we ignore the underlying absurdities.

It was my good friend, Ken Laker, who taught me to be binary about behavioral issues. Just like Mr. Gore, Ken also literally changed my life. Right after I completed my Masters degree in Electrical Engineering, Ken offered me a job and then immediately turned around and advised me not to take it. He told me that once I start earning $30K a year, I would find it very difficult to go back to a $10K a year life as a poor graduate student. Therefore, since I had an offer from Stanford University for a teaching assistantship while pursuing a Ph.D. in Electrical Engineering, I should take that offer from Stanford over the job that he was offering. I took his advice and had the time of my life at Stanford, where I also met the amazing Jaine, who later became my wife.

Ken was a jolly, well-proportioned man, who laughed with his entire body, which made all his jowls shake. Three years later, I saw him again after my Ph. D. and he once again made me a job offer. I didn't accept his job offer this time because I didn't think that the new research organization that he was working in would last long. But he was his same jolly, well-proportioned self and he remained so for the next few times that I saw him over the years.

But later, in the mid nineties, I ran into Ken once again and was stupefied to discover that he had lost almost half his weight and was looking extremely trim. I hugged him and said, "Wow, Ken, you look great! What happened?"

He said, "You mean, my weight?"

I said, "Yes." Ken replied,

> "Well, I finally ran into a doctor who turned it around for me. I was feeling really sick and went to this doctor, who ran some tests and said,
>
> 'Ken, you are going to die if you don't reduce your weight. I want you to eat nothing until Friday and come back to my office for a follow-up check.'
>
> Until then I had been on several diets and none of them had worked. The diets were all about eating less of this and more of that and I never could stick to them and I always reverted back to my old habits. But, when this doctor told me to eat nothing, I understood that very well. I knew what ZERO meant.
>
> I followed his advice and then went back on Friday. He did some more tests and then said, 'OK, now you can drink two glasses of orange juice every day until Tuesday. Come back on Tuesday and we'll do another check.' And the doctor slowly rebuilt my diet from scratch and I'm now eating healthy.
>
> I owe him my life."

Since then, I've become a true believer in the power of ZERO, thanks to Ken! I do not believe in gradualism or moderation in the face of absurdities. Once I consciously understood that consuming animal products was senseless, I dropped them instantly. And a month after I turned vegan, I felt this enormous sense of guilt lift off my shoulders and relief wash over me until tears came into my eyes. I must have subconsciously known that my consumption of milk, eggs and leather was shameful and I had been suppressing that knowledge all along. Later I asked my uncle what our grandparents did with the dairy cows that stopped giving milk and reached old age in our farms back in the village. He said that they used to sell the cows to their Muslim neighbors for you know what.

Therefore, I believe that the most effective step that I can take in an urban setting to promote the Butterfly culture is to switch to an organic, vegan diet. Besides, no one can force me to eat animals or pesticides against my will. Moderation campaigns such as Meatless Mondays do serve to highlight the undesirability of meat consumption, but they also mislead people into believing that they are doing good when they are actually doing more harm. For instance, dairy cheese is known to be more carbon intensive than certain meats like pork and chicken[17] and on Meatless Mondays, people tend to splurge on cheese.

It is easy to justify our consumption of meat, dairy and eggs with rationalizations such as, "Well, that's only 5% of my daily consumption and therefore, that's just a small blip in the bigger scheme of things." But, when we accept that those are foods, then when we celebrate an occasion, we tend to order buckets and buckets of the stuff because we've rationalized our continued addiction to those substances. Then the advertisers and marketers work over us to persuade us to treat every day as an occasion and to live life plentifully. And then nothing really changes in the world around us.

The so-called "sustainable meats" typically involve reverting back to the old way of raising animals, on bucolic farm settings. But factory farms arose in the first place in order to reduce the resources needed for producing animals. The corporations that run these factory farms employ scientists with Ph.D.s who systematically determine how to optimize production. These scientists even studied how soon the calves had to be separated from their mothers in order to optimize milk production. The answer, within 1-3 days, for otherwise the mother gets attached to her calf, and starts bleating and refusing food, thereby diminishing her milk production[18]. Besides, traditional farms waste food by allowing the animals room to frolic and expend energy. It is far more efficient to have them stand still, as in factory farms, to maximize the conversion of the food eaten by the animals into meat and dairy. Prof. David Pimentel estimates that grass-fed beef requires twice as much water to raise as factory-farmed beef[19]. Twice as much water implies twice as much land for the same amount of meat and consequently, if everyone switched to such meats, there would be no more forests left on planet Earth. Of course, we can price the meats higher so that it is out of reach of the masses, but that would be construed as an elitist plot and tremendous pressure would be brought to bear on politicians to subsidize the meat to the same per-capita end consumption level as it is today.

Finally, the argument that we still need to raise livestock in order to generate the manure that we need for our organic produce falls flat on its face upon closer scrutiny. I asked an ecologist friend of mine, "Surely, we know of better ways to compost biological matter than by feeding it to a cow and collecting the manure at its rear end?" and he replied, "Whoa, we would need to radically alter societies at the village level if you are asking them to switch to vermi-composting[20]!" But, even if rural communities take a while to change to such simpler and more efficient processes, it is up to us in urban communities to stop enslaving livestock with the excuse that we need them for their manure. With worms doing all the composting that we need instead of cattle, we won't be tempted to turn the worms into steaks and hamburgers and return to our Caterpillar ways.

Habits change and new taboos get created all the time. We no longer burn slaves at the stake for our night-time lighting as the Roman nobles used to do two thousand years ago[21]. We no longer lynch people in public because they looked "uppity" and happened to have a different skin color[22]. We no longer practice cannibalism with the rationale that our dinner came from a different tribe and therefore, deserved to die anyway[23]. Likewise, our animal enslaving practices have to become taboo in the Butterfly culture. This is already happening among the Miglets: a note on the "Occupy Wall Street[24]" web site recommends that well-wishers purchase vegan pizza for the protestors so that, and I quote, "everybody can eat it."

This idea of low-footprint, conscious consumption extends to material goods as well. With the technologies that we have developed over the years, there is so much chemical pollution that accompanies each product that is manufactured. The documentary,

"The Human Experiment[25]," details over 80,000 chemicals that our industrial and household products have unleashed on the environment and that we don't really understand the biological implications of most of these chemicals. While organic veganism minimizes our ingestion of these chemicals in our foods, it is through reducing our conspicuous consumption of "stuff" that we minimize the introduction of these chemicals into the environment in the first place.

7.3 The Creation of Abundance

The second most effective step is to go local, to strengthen local communities, to build up the Transition Towns[26] and to grow our organic produce locally. Urban areas constitute land that we've appropriated from Nature for human use and it behooves us to make the most of that land, to grow fruits, nuts and vegetables in all the sidewalks, backyards, front yards and parks. If Amory Lovins can grow bananas in Colorado[27], surely we ought to be able to grow most of what we eat locally? This would allow us to return most of the rural land, especially the land that we're currently using for livestock production, back to Nature to regenerate forests and to heal the biosphere. This is the "Garden World" vision of Doug Carmichael[28].

Making maximum use of the urban land that we've appropriated also means that we would tap into the solar, wind and geothermal energy within these urban areas to meet our energy requirements. The energy infrastructure of our cities would mimic the communications infrastructure of the internet, with a distributed, robust, non-hierarchical structure that is difficult to disrupt. It is only when we cease to rely on remote, concentrated energy production that the hierarchical Caterpillar culture can dissolve and the true Butterfly can emerge en masse.

The Butterfly culture is about the creation of abundance. In contrast, in the Caterpillar culture, a substance that is abundant would be considered worthless and it is only by making it scarce would it be rendered valuable. I recall that around thirty years ago, when Perrier first introduced the concept of bottled water in

America[29], it was met with derision. When tap water is nearly free, clean and abundant, why would anyone pay money to buy bottled water from a French company? Yet the young urban professionals of that time bought it because the company cleverly sowed doubts about the cleanliness of tap water. The bottled water industry grew from this small beginning into the 100 billion dollar a year behemoth today[30], by creating both the perceived and the real scarcity of clean, drinking water and by manufacturing demand. Thus, the Caterpillar culture did the opposite: it took something that was abundant and turned it into a scarce commodity, by physically, chemically and psychologically polluting it, while positioning the industry as the sole means for providing a cleaned version.

Now imagine taking something so polluted and cleaning it up so that it becomes free and abundant. And imagine doing this with volunteers who are willing to work on the project because they believe in the end goal of abundant, free water for all. So much so that the volunteers bring in not just their sweat equity, but procure all the equipment needed to complete that task. That is the Butterfly culture. Things get done, even big things get done, because enough people believe in the end goal. Things get done because someone has inspired enough people to work on it, because it makes sense within the overall purpose of the Butterfly, to regenerate and celebrate Life.

This is the gift Economy of the Butterfly. In the Butterfly economy, food is grown in abundance so that no one goes hungry. Shelters are built as and when needed just as the Amish do, not because the construction workers get paid to do the job, but because they are celebrating that rite of passage for a young couple. Everybody goes to work because they truly believe in the work that they are doing. Everybody would literally pay somebody to do the work that they are doing. Absenteeism is a rarity in such a culture because it is fundamentally driven by Love and not by Fear.

And clean water flows freely.

If you need a software operating system, just write the initial code and make it open source so that your friends can help you perfect it. And Linux is born[31]. If you need an encyclopedia, just ask. And Wikipedia is born[32]. If you need a web site, just ask. And CharityFocus will do it for you[33]. If you need help with homework, just ask. And Khan Academy is born[34]. If you need something, anything, just ask. And, if enough people believe in why you need that something, anything, you will get it.

For free.

As we are slowly beginning to discover, money is not a good motivator for creative work. Prof. Dan Pink of Harvard points out that it is autonomy, mastery and purpose that are much stronger motivators than money[35]. The best thing we can do with money is to create an environment where money becomes irrelevant. Then, give people the freedom to create. But Sir Ken Robinson says that we are educating children out of their creative capacities at the moment[36], in our Caterpillar culture. We're running national education systems where mistakes are the worst thing that children can make, but it is mistakes and group learning that foster creativity. In our current education system, we are chipping away at the creativity of the individual in order to chisel out a Lego piece that can be plugged into the vast pyramid that is the global Caterpillar economy.

The Chilean economist, Manfred Max-Neef, summarizes the principles of what I call Butterfly Economics, using five postulates and one fundamental value principle[37]:

1. The economy is to serve the people and not the people to serve the economy
2. Development is about people, not about objects.
3. Growth is not the same as development and development does not necessarily require growth.
4. No economy is possible in the absence of ecosystem services.
5. The economy is a subsystem of a larger finite system, the Biosphere, and hence permanent growth is impossible.

And the fundamental value principle to sustain the economy is that no economic interest, under any circumstance, can be above the reverence of Life.

Within Butterfly Economics, development is about creative possibilities, which is infinite, while growth is about quantitative accumulation which is necessarily finite. There are absolutely no limits to development in a Butterfly economy because its objective function is the creativity of society, not the totality of goods produced.

In the Butterfly culture, everyone is working towards a common purpose, the regeneration of Life, and there is this deep meaning to human lives. The third pillar of true happiness, Purpose, is built in to our lives. In his very spiritual book, "Man's Search for Meaning[38]," Viktor Frankl writes about how he survived the Holocaust because he had found a meaning for his Life; he wanted to find his wife. His love for his wife let him forget the trials and tribulations of concentration camp and let him forgive his captors for causing them. Therefore, it is Compassion, Detachment and Purpose that make up the three pillars of true Happiness.

Ideally, the actions of a Butterfly are performed without any ego, without any desires and while forsaking the fruits of those actions. For it is wisdom that defines a Butterfly, not knowledge. In the Butterfly culture, knowledge is available at everyone's fingertips and therefore, people are constantly striving to be wise. And a truly wise being can never be boastful, egotistical and will forever be steeped in humility and wonderment towards Nature. Humans have good reason to be humble as a recent article in the New York Times detailed how animals and birds outperformed human beings in various tests, including social interactions (chimpanzees), botany (sheep), probability and statistics (pigeons) and common sense (crows)[39]. The test with the pigeons was particularly striking as it involved a famous probability dilemma that I was introduced to during my years at Stanford. It is based on the old game show, "Let's Make a Deal," where Monty Hall was the host. In the game show, the contestant is asked to choose from three doors. Behind one of the doors is a spectacular prize, e.g., a car, and behind each of the other two doors is an ordinary prize, e.g., a bicycle. Once the contestant makes his choice of the three doors, Monty Hall reveals the bicycle behind one of the other two doors and then gives the contestant the opportunity to switch his choice to the other unopened door. Most of the contestants on the game show did not switch.

This is contrary to the theory of expectations. When the contestant made the initial choice, he had a 1/3 probability that he picked the car. The probability that the car was behind one of the other two doors was 2/3. If, instead of revealing the door with the bicycle, Monty Hall had offered to the contestant the chance to switch between the door he picked and both of the other doors, surely the contestant would have switched? That is, if he gets to keep the car if it is behind either of those other two doors. And that is precisely the option the contestant is given when Monty offers the chance to switch once he reveals the bicycle. In laboratory tests, only one-third of human subjects switched. But a whopping 96% of pigeons switched in a version of the Monty Hall dilemma involving pecking keys and "mixed grain" rewards.

Pigeons 1, Humans 0.

However, whether animals and birds are smarter than humans in cognition tests - that humans devise - is beside the point. Indeed, during the recent earthquake in Virginia, many animals in the Washington National Zoo appeared to have anticipated the earthquake well before it happened[40]. Therefore, if animals devise tests for humans, they might invariably find us dumb and lacking in basic, common sense. As the biologist Richard Dawkins once pointed out, every species on Earth today is at the same level of the evolutionary tree. We are all truly cousins, the human, the tiger, the bat, the shark, the oak tree and the fungus. And a truly wise species wouldn't be devising tests to judge others and wouldn't even be calling itself "Homo Sapiens," the Wise Man.

For have you ever met a truly wise man who's constantly boasting that he is wise?

As such, the Butterfly culture is terrifying for the human ego, because it would be impossible to persuade enough people to voluntarily build a monument to its arrogance and stupidity. Why would volunteers build a 27 story, one billion dollar monstrosity to house an individual and his five family members, with a helipad on the roof just so that they won't ever have to set foot on the ground?

The 27-story Ambani family home, Antilia[41], would have never been built in Mumbai, India, in the Butterfly economy.

Fig. 6. *The plight of dairy calves speaks to the healing that we need to accomplish.*

8. Climate Healers

"Tell them and they will forget. Show them and they may remember. Involve them and they will understand" - Ancient Chinese Proverb, attributed to Confucius.

In our circle of family and friends, the 27-story, 400,000 square foot, Ambani family home, Antilia, in Mumbai, India, is a source of much derision. It cost a billion dollars to build and furnish, in a show of wealth fit for a modern day Maharajah. It is the home for a family of 5 individuals and 500 servants in a ratio that is a fitting parallel to the 1% vs. 99% debate that is raging throughout the world today. But, in our circle, Antilia is viewed as a monument to one man's hubris, a monstrosity of opulence in the midst of so much poverty, built in a nation where 48% of its children below the age of five are malnourished, a truly sickening misallocation of precious resources[1].

But it is very hard for us to step back and recognize that our modern industrial civilization is just one gigantic Antilia, from the viewpoint of our fellow Earthlings. As far as the Tiger is concerned, there are very few humans who are not Ambanis or the Ambani servants. Almost everything we do, including most of our philanthropic activities are geared towards improving the well-being of ourselves and our fellow humans, with little to no thought given to the well-being of other species. This is precisely like Mukesh Ambani showering his wealth on himself and his family in the midst of so much poverty, while possibly doling out raises to his servants once in a while. Can you imagine Bill and Melinda Gates and Warren Buffett to be Ambanis, who just happen to be a little more generous to the servants, the "99%"? Well, there are very few non-Ambanis among us from the viewpoint of the tiger. And our human enterprise looks like a monument to one species's hubris, a monstrosity of opulence that covers two-thirds of the land area of the planet in the midst of so much poverty, built in a biosphere where 97% of the tigers are not just malnourished, but simply dead. It is a truly sickening misallocation of precious planetary resources.

When I began this journey four years ago, I was at a loss to know where to start, but I instinctively knew that our Caterpillar culture has got to be dismantled. A centralized, fossil fuel energy infrastructure is a deadly combination with ubiquitous, distributed communications as that allows large corporations to reach out, influence everyone and turn us all into Caterpillars. Therefore, it is imperative that the energy infrastructure of the world be changed to promote the Butterfly culture.

Climate Healers[2] is a US 501(c)3 non-profit corporation set up to promote the Butterfly culture, the culture of Life. Its stated objective is to facilitate reforestation while minimizing fuel use for cooking and lighting systems in low income neighborhoods throughout the world. Unofficially, it grew out of my frustrations with just talking about climate change through the Climate Project presentations and not doing much about it. In late 2007, I became frustrated enough that I sought the advice of my dear friend, Juan Jover, and it was he who told me to start a non-profit in the clean energy sector.

And the Lighting Project[3] was born.

8.1 The Lighting Project

The Lighting Project was started to address just the lighting energy needs of the rural poor in a distributed manner so that they don't have to depend upon kerosene and other fuels for that purpose. When Thomas Edison invented the incandescent lamp, he famously said that "we will make electricity so cheap that only the rich will burn candles[4]." But, as of 2000, there were more non-electrified households than the total number of households during Edison's time.

The underlying premise of the Lighting Project was very simple. If we take the number of people who do not have access to electricity worldwide (about 300 million households) and the estimated amount of kerosene that is used for lighting worldwide (about 20 billion liters per year), then a strong business case can be made that all these households can be supplied with two free solar LED lights at $10 per light using the carbon credits from the kerosene that they won't be burning. We calculated that it would be a flourishing, sustainable enterprise if carbon credits fetch at least $11 per metric ton. And, off we went.

I believe in trying things out and plunged in head first by identifying a Non Governmental Organization (NGO) in India, ordering 1000 solar LED lights from SunNight Solar[5] in Houston and having them shipped to India to begin a pilot effort. Our NGO partner is the venerable Foundation for Ecological Security[6] (FES), which is a non-profit organization dedicated to regenerating forests and as such, works in remote areas of India. FES originated from the campuses of the National Dairy Development Board in Anand, Gujarat, India, which further cemented the relationship between dairy production and deforestation in my mind and drove me to veganism. We identified two villages, Karech in the Udaipur district of Rajasthan and Hadagori in the Dhenkanal district of Orissa, for the pilot implementation. FES had been working for several years in these districts and had a good rapport with these villagers. The village in Rajasthan is near the Kumbalgarh Wildlife Sanctuary[7], which is the main green patch that is preventing the Thar Desert, which actually originates in the Sahara in Africa, from sweeping into New Delhi. Karech gets relatively little rain throughout the year, while the village in Orissa experiences the South West monsoon season during summer. Therefore, we thought that if we can get the pilot project to be successful in these two villages, then we could probably get it to work almost anywhere in the world.

After shipping the solar lights to India, I flew there to be in the villages during the distribution. But, as luck would have it, the shipment got stuck at Delhi airport because FES's license to import goods, called the Import Export Code (IEC), had expired. And I was about to get a lesson in the sophistication of the corruption that rots the Indian Government's bureaucracy.

India has always had corruption in the bureaucracy, but when I was living there in the sixties and seventies, the practitioners were quite crude about it. At that time, you could get things done without having to pay bribes as most of the bureaucrats were afraid and even ashamed to be seen as corrupt. But nowadays, the corruption appears to have become truly endemic, as if there is a parallel, shadow taxing mechanism that funnels wealth directly to the bureaucracy and to the politicians. This is what Anna Hazare and the Miglets are agitating about in India today.

Fortunately, all the interactions with the bureaucracy occurred through FES, specifically a young employee of FES, who appeared to be a novice at wading through the bureaucratic thickets. He looked up the process for renewing the IEC, downloaded the necessary forms, filled them up and sent them in.

And nothing happened.

As the first direct application simply disappeared, he sent in a second application for the renewal, followed up on the phone, and got back a response that the application forms were printed off the internet and were therefore, invalid. It seemed that the real application forms had to be requested and mailed from the office in New Delhi after paying a nominal fee and only then will the application be considered valid. He did that and there was still no response. Then, a few weeks and many phone calls later, an official suggested that we hire a special agent in Delhi, who understands the application process intimately. That we shouldn't be trying to renew the IEC code without the help of such a "professional."

Meanwhile, for every day that the shipment was held up in Delhi airport, there was a hefty rental charge being added to the shipping bill.

Finally, FES hired a special agent in New Delhi as recommended and had the IEC code renewed roughly three months after the shipment of solar lights reached Delhi airport. The lights were distributed to the villagers by FES personnel in my absence.

When I first visited Karech and Hadagori in March of 2008, I went with the FES personnel, but empty-handed with just the idea of the project and no solar lights. We explained to the villagers what we were planning to do, but within the first half hour of my visit to Karech, the business model for the Lighting Project fell apart. It turned out that these villagers only bought their kerosene from the government ration shop, which was several miles away from the village and required a full day trip for the women. And quite often, they would make that trip only to discover that the ration shop was closed. But kerosene cost just Rs. 10/liter in the ration shop whereas the open market price was double that amount giving the ration shop operator an incentive to sell their unclaimed quota on the open market and make some extra money. Therefore, Kerosene was a precious commodity for the villagers.

And for the most part, they weren't using it for lighting at all.

They were procuring at most 3 liters of kerosene per household per month and most of it was really being used to start cooking fires. One villager asked me, "Why do you want to change the night? God gave us night so that we can go to sleep!" This was a very humbling question for me. But, since we were giving away the solar lights for free, the villagers were keen to try them out.

Meanwhile, we discovered how much wood they were burning in the village for cooking and it was shocking not only to me, but to the FES personnel who had accompanied me to the village. Since Karech was close to the sanctuary, the villagers were using as much as 20kg per household per day and even an extra 10kg at times to sell to neighboring villages that were further from the sanctuary. They said that once the government built roads into the village three years ago, it had become easy to transport the firewood and other forest products to the neighboring villages and that this has increased their income in the village.

I made some quick mental calculations, determined that a similar carbon credit mechanism would be easily workable for solar cook stoves, and asked the villagers if they would be willing to use them if we could make them available free of charge. They said, "Show us how to make Jowar rotis and we'll use them." Jowar rotis are thick flatbreads made from a type of sorghum and they require a lot of energy to cook. Along with Corn rotis, they form the staple diet in these village communities. I promised the villagers that we would ensure that the solar cook stoves can make these rotis and left.

We got very similar reactions during my visit to Hadagori, Orissa as well. Once again, lighting was not viewed as a major requirement by the villagers, but they appreciated the fact that I had come all the way from America to help them with their energy needs. One woman actually berated the menfolk that they ought to be ashamed of themselves for drinking so much alcohol when this stranger had come all the way to help. It seemed that alcoholism was rampant in the village among the men, but the women were all teetotalers. This is a source of considerable friction between the genders as the men waste the meager resources of the household on their addiction.

We discussed the cooking needs of the villagers and they asked us to show them how to cook rice and lentils on the solar cook stoves if they were to use them.

Six months passed by before I returned to the villages. The solar lights had been distributed in the villages three months prior to my September 2008 visit and the monsoon season had just ended. My first stop was in Karech, Rajasthan and I was almost mobbed by the same village women who had been so shy and reticent during my first visit. Apparently, within three months of the distribution, the solar lights had become the most important possession of every household in the village, especially among the women.

There were a couple of reasons for the popularity of the lights. Since the lights were distributed, thefts had become non-existent in the village. In fact, the villagers chased away a thief who tried to steal a buffalo, as everyone ran out with their lights upon hearing the commotion and the whole village lit up like a city, frightening the would-be thief. Secondly, and more importantly, the previous three months was snake-bite season in the village. The women go out into the forest to do their ablutions early in the morning before the men wake up and would get bitten by snakes, especially during those months. Every year, 20-30 women used to get bitten by snakes and 2-3 women used to die from snake bites. And that year, in 2008, they had ZERO snake bites in the village because the women used the solar lights when they went into the forest and could see the snakes before they stepped on them.

The main complaint they had about the lights was that the switches were flimsy and we replaced 21 of the 480 lights in Karech for that reason. We left Karech loaded with gifts of delicious, crunchy giant cucumbers, and we ate one on the return trip during our lunch break.

My trip to Hadagori, Orissa, fell two days after the Mahanadi River had breached the bridge between Bhubaneshwar and Cuttack and cut off the road in what was the worst flooding in Orissa since 1982[8]. The monsoon rains had stopped but the water was everywhere. Fortunately, the Mahanadi River receded on the day of my arrival allowing us to travel to Hadagori from Bhubaneshwar. We managed to visit two of the hamlets in Hadagori, while the other two hamlets could not be reached due to the depth of the mud covering the roadway. In both these hamlets, the men complained that the women were not letting them use the solar lights to play cards at night, but the women said that they allowed the men to use the lights after the children went to sleep around 10pm, but that the men were too drunk by then to do much with the lights. The women used one light for cooking and the other for the kids to study, and the men's gambling activities were of lesser priority to them.

I heard the snake bite story in Hadagori as well. It was shocking to me that if this was happening in both Rajasthan and Orissa, then tens of millions of women throughout India are literally risking their lives every day while doing their most basic biological functions! In Hadagori, the villagers were also using the lights to chase away elephant herds that wander into their crop lands. And they had found a way to charge the lights even during the monsoon season by wrapping them in transparent plastic covers and leaving them out in the diffused light. They rationed the use of the lights when it is cloudy outside and made them last for at least 4 days before the batteries ran out. The solar lights had become so popular that neighboring villagers were very envious of them and were offering princely sums, as much as Rs. 2000 for the lights, but the villagers had not sold a single light as they wanted us to continue our project with them. Our FES partners said that they are becoming known as the "Torch people" among the villagers whereas previously, they had been known as the "Forest people."

Though the original business model of the Lighting Project was not workable, the solar light intervention taught us many lessons. The lights helped the villagers expand their daytime and improve their productivity. On the flip side, the forest was already overexploited, especially in Rajasthan, and the use of the lights further accelerated the exploitation of the forest. We heard reports that the women were making plates out of leaves at night and some trees were completely stripped of their leaves in the process. The income of the villagers increased by about 15% due to the intervention, but at the expense of the forest.

The intervention was done collectively, with the entire village participating in the project. This seemed to enhance the community experience as the lights were used during festivals and weddings to light up the night for dancing in the village commons, with every household contributing one light to the occasion.

FES, our NGO partner, was key to the success of the intervention as they were present on the ground to replace defective lights and to ensure the satisfaction of the villagers. Also, it was FES's rapport with the villagers that gained us an audience in the first place.

Best of all, using the solar lights did not require any change in the daily routines of the villagers as they charge the lights on their rooftops or porches unattended during the day, and use the lights at night once they return from their field work. And we were about to find out how important this is to the success of any intervention.

8.2 The "Namaste" Solar Cook Stove

Cooking was clearly the most dominant energy consuming activity
in the village households. Even if all the kerosene that the villagers
procured had been used for lighting and none for kindling cooking
fires, 3 liters of kerosene per household per month works out to at
most 0.1kg of kerosene per household per day. But the villagers
were using 10-20kg of firewood per household per day, which is 50
to 100 times the carbon content of the fuel used for lighting. The
Lighting Project became Climate Healers.

Worldwide, nearly 3 billion people use some form of biomass for
cooking, burning up an estimated 1.5 billion tons of firewood
annually[9]. That is the amount of wood that would grow on about a
billion acres of land in a temperate region, each year, if left alone.
This statistic is a searing indictment of Catapilism, the distorted
version of Capitalism that is dominating the global economic
system today. Massive investments are being made in the endless
production of entertainment gadgets for the rich, but very few
investments are being made to satisfy the basic needs of the poor.
Since Life is dying in the process, it couldn't be enlightened
self-interest that is driving the economic engines of the world, but
its very opposite.

Instead of cooking on a three-stone fire, if the villagers used more
efficient cook stoves, their firewood use could diminish, but there
were several problems with that approach. Firstly, such efficient
stoves depend upon the firewood being of the right size and type
and if it is not, the efficient cook stove could belch far worse
smoke than the traditional approach. Even in these remote areas,
firewood smoke contains industrial effluents since the trees have
been filtering our industrial pollution from the rain water and
storing it in their trunks over the years. And efficient stoves require
a steady supply of oxygen in their air intake, which becomes a
problem in indoor settings, especially if the air intake path is
clogged with ash.

Secondly, energy use among the poor will necessarily increase as
they develop and treating the cooking problem as just a stove

efficiency problem leaves the source of their energy the same - firewood. This means that we would eventually need to address the energy source issue even if we provide them with technologies to burn firewood efficiently. Therefore, we decided to help them tap into an alternate source of energy that is falling plentifully on their heads - solar.

Faced with the task of supplying a solar stove that can cook rotis made from various local grains such as Bajra, Jowar and corn, in addition to boiling rice, lentils and vegetables, we hired an engineering firm in Ahmedabad, India to either procure or design such a solar cook stove along with an electronic mechanism to meter its use. The meter was to measure the number of hours that the cook stove gets used so that the user can be compensated at a proposed Rs. 2 per hour as an incentive, which can be financed through a carbon credit mechanism. Cooking rotis is a little tricky: if the time taken to cook the roti is too long, then it becomes hard and inedible and if it is too fast, then the insides would be uncooked even as the outer skin gets charred. There is a Goldilocks range for the power requirement depending upon the grain and thickness of the roti, and it was around 1 kW. But every existing solar cook stove that could supply 1 kW at its cooking surface was huge and we couldn't stand near it to flip the rotis. These stoves were all of the parabolic kind that concentrate the solar energy falling on their reflectors to the center, but the energy leakage was so high that more than half that energy wound up cooking the cook.

The "Namaste" Solar Cook Stove design mainly had to improve the efficiency of existing solar cook stove designs, while engineering it to be foldable and portable so that the user can put it away in a corner of their hut after use. The efficiency was improved by adding a shielded metal skirt at the focal point of the parabola so that the thermal energy is trapped by convection currents. The design was made foldable in the same manner as in the "Butterfly" cook stove, by splitting the parabola into two separate halves. And we added wheels to make it portable. We called it the "Namaste" stove since in its folded state, it resembled the Namaste gesture of greeting that people in India use with folded hands. Folded hands resemble the flame of a candle and the gesture is used to acknowledge the Light or the Atman within another.

The first prototype was deployed in Karech, Rajasthan, in September of 2009 and after incorporating some feedback from the deployment, 5 more cook stoves were deployed, four in Rajasthan and one in Hadagori, Orissa in January of 2010. More detailed feedback was collected from the villagers six months later. Unlike the solar lights for which the feedback was overwhelmingly positive, the feedback in the case of the "Namaste" solar cook stove was overwhelmingly negative. Here is a compilation of the villagers' reactions to the stoves, gathered through my personal visits and through FES personnel[10]:

1. The solar cook stove is difficult to use in hilly terrain without leveling the ground.
2. It is uncomfortable to stand facing the sun to cook without an umbrella shade.
3. The cook stove is too tall for the women who actually prefer to sit and cook.
4. The solar cook stove requires constant adjusting which makes it inconvenient to use.
5. It is difficult to use the solar cook stove within the normal routine of the village women, since they are rarely in or around the house during the day. One villager even said to me, in effect, "Why did you do the cooker differently from the solar light? With the solar light, we left it on top of the hut and went to work. When we came back in the evening, it was ready to use. The cooker didn't work the same way".
6. The solar cook stove is difficult to use with a child in arm or a youngster in tow as it requires the user to stand and stretch to reach the cooking surface.
7. The solar cook stove is unstable and topples over in windy conditions, spilling the food.
8. The solar cook stove can only be used outdoors and during day time when the woman has various unending chores to tend that center around her children and animals.
9. Since the solar cook stove was only given to a few families in

the villages, the recipients felt left out of the social activity of gathering firewood in the forest in the company of their friends.

Though most of this feedback is applicable to any parabolic solar stove design, the net result was that over time, the "Namaste" solar cook stoves were moth-balled in the villages. This experience was so unlike that with the solar lights which continue to be treated as the prized possession of each household. In that context, the villagers' feedback has been truly humbling. In the prototype deployment, we did not install meters to measure the stove usage and provide incentives for the users, but it is unlikely that payments on the order of Rs. 2 per hour of usage would have improved the feedback substantially. This is especially because the Indian Government has instituted the National Rural Employment Guarantee Act[11] (NREGA) program in these villages and as a result, households are earning an additional Rs. 100 per day which dwarfs our proposed incentives.

8.3 Miglets to the Rescue

Just before my trip to India for the "Namaste" cook stove evaluation, I received a phone call from Prof. Rajagopal (Raj) of the University of Iowa asking if Climate Healers and FES could host some students from the university during their winter break in Rajasthan to work on the solar cook stove project. The students would be part of the Iowa Winterim[12] program, an innovative one credit project course that gives students the opportunity to visit India and do some social work while experiencing a cultural immersion. I readily agreed and proposed that the students could help with the widespread deployment of the "Namaste" solar cook stove in the village as their project. I returned from my India trip knowing that there wasn't going to be any such large scale deployment.

I wrote up a report on the status of the cook stove deployment in August of 2010 and sent it to all the patrons at Climate Healers and to Raj, expecting that the proposed Winterim program with the Iowa students was probably going to get cancelled. I didn't hear back from Raj until early December when I received a short e-mail from him stating that 9 students and a faculty member will be arriving in New Delhi on Dec. 27 to help with the project. It was a wonderful surprise and a major turning point for Climate Healers.

The nine students and the professor from Iowa who came to Rajasthan were the perfect team to do an assessment of the cooking needs of the women in the villages and to redefine the solar cook stove project. They were three chemical engineers (Jackie, Ben and Ethan), three mechanical engineers (Matt, Eric and Brianne), one anthropologist (Billy), a journalist (Abbey) and an environmental engineer (Josh). They were typical kids from the American heartland, authentic, hard working, fun loving and simply amazing. Most of them had never gone far beyond Iowa in their lives, but they were the best Miglets that we could have asked for. The professor who accompanied them was Prof. UdayKumar, from the mechanical engineering department at the University, once again a perfect match for the project. The liberal arts majors worked on a crisp questionnaire to understand the needs of the women and the Miglets split up into three teams, each with a local interpreter and a woman student so that the teams could establish good rapport with the women in the villages. And it worked beautifully.

Here's how Abbey, the journalist, described her experience[13]: "In a small village called Karech in Rajasthan, India, Rayakbai walks 3 kilometers uphill every day to collect firewood for her family. She says she has been doing this since she was 7 years old, just old enough to help her mother. But within the last five years, her work has become increasingly difficult. She remembers when the trees were prevalent in her village and walking to get firewood only took a small portion of her day. However now, most of the timber is gone and she has to walk farther to collect the wood her family needs, taking three to four hours to complete the task. When she returns home, she cooks two meals per day. The smoke burns her eyes while she cooks in her small, unventilated kitchen, but she has no choice because her family of five has to eat. ...

Rayakbai had lived in the village her entire life. Her home was nicer than most, with stone walls and thick branches for a roof. As we walked in to talk with her, her sons brought a bed made of thick wooden legs and tightly woven rope for us. Sailesh motioned for me to sit and as I did, the rope sank lower and lower towards the ground. I winced. The last thing I would want to do is break one of the only belongings this family owned.

Outside, the January sun beat down at us at 70 degrees Fahrenheit, but I shivered in the shadows of Rayakbai's living room. Goats walked in and out of the room, bringing the smell of farm and manure in with them. We began to ask questions, like "What would your perfect stove be like?" and "What do the men do during the day?" Soon our group discovered that there would be many more constraints than we imagined if we were to design a proper solar cooker. The men drank during the hot afternoons, leaving the women with most of the work. If we gave them a woodless stove, the men would surely just give them more work to do despite the gained free time.

Their young children sat staring at our blonde hair with their large, brown eyes. When the women were alone, they smiled, with red lipstick on their mouths, laughing with us like old friends. When men surrounded them, they turned their backs to us and would not speak. Sailesh turned to me, noticing the makeup, and said, "See, they want to be like us. They don't want to live like this."

Being the journalist of the group, I took control of constructing a questionnaire to give to the women to better understand their cooking habits. My group wanted to ask questions that were direct and simple to interpret because the villagers in Karech spoke Hindi, mixed with some native dialect, so we would be using interpreters. As we got to talking to the women though, I realized I could communicate easily with them without even knowing the language. Their laughter, sorrow, and concerns were contagious, and bled through the words so I still understood what they were trying to say. Her smile shone through as she told us what she knew. She lifted up her skirt, showing a large bump behind her knee where she had fallen recently while she was collecting wood. Things needed to change, for her, and for the forest.

Carbon Dharma

Raykabai's 8-year-old daughter had just started to accompany her to the forest daily to pick up 44 pounds of firewood. They both carried separate loads on top of their heads down the rocky roads, with no shoes. Her feet were covered in dirt and blisters, her long toenails running rampant underneath her skinny legs.

The women needed something to change. They were losing their forest every day, and they knew it would not come back instantly. Through our questionnaire, we discovered they saw the connection between the plentiful forests bringing rain to help their fields. But they would not change unless it was easy. Their families had been there for nearly 200 years, the lifestyles rarely changing".

As we were conducting the survey, Uday got excited about going back to Iowa and working on the design of a stored energy solar cook stove with the Mechanical Engineering Design Project (MEDP) class during the Spring Semester of 2011. He said that he had been dreading that class as it usually involves students working on minor modifications of mechanical shafts and gears for local corporations like John Deere, but he would now suggest to the class that they design a stored energy solar cook stove that meets the villagers' needs as a joint class project. That class of 30 students worked on the project along with four students from the University of California, Berkeley and I've never met a more enthusiastic group of students in my life. They were truly inspired to work on the project and they had so many ideas, but they had to settle on implementing just one prototype due to resource and time constraints. Unfortunately, that prototype didn't perform as expected for various reasons, which Uday and a fresh batch of students are sorting through.

During the next semester, Climate Healers is issuing a challenge[14] to schools and universities in the US and India to work on the design of a stored energy solar cook stove that Rayakbai would love to use. Four Miglets, Billy, Sita, Shaanika and Rama have been sending letters out to schools and universities soliciting their participation. For example, here's the letter that Billy has been sending out:

"Hello, I am Billy Davies, a senior at the University of Iowa and intern for Climate Healers, a non-profit organization that partners with student groups, universities, and NGOs and rural villages throughout the world in order to curb the effects of climate change by regenerating forests.

This year, Climate Healers has launched a challenge open to any engineering club, student organization, or university groups in which teams will work to create a design for a solar-powered cook stove that will be deployed in selected rural villages in India.

I wanted to share with you my story of how this great initiative got started and show you how this challenge is the start of a collaborative effort to undo the damage caused by deforestation and to improve the lives of millions and perhaps, billions:

Last winter, I had the privilege to be a part of an excursion to the city of Udaipur in Rajasthan, India with eight other University of Iowa students, our supervising professor, H.S. Udaykumar, and Climate Healers's Executive Director, Sailesh Rao. Our research team of six engineers, one anthropologist (myself), one journalist, and one environmental scientist set out into the villages of Mewar and Karech to examine the effects of wood cutting and burning on the villagers, how the intensity of fuel wood use has increased over the last five years, and ultimately examine why previous efforts by Climate Healers to deploy a suitable solar cooker prototype in the field were unsuccessful.

Our time in India was short, but the experience is something that none of us will forget. Even in our short sojourn in the North, we saw that India is a place with a rich heritage and of great cultural and ecological diversity. We were also awed by the after effects of years of imperialism and environmental degradation that have left the country's people and its ecosystems in distress.

Through face to face interviews of villagers and a lot of help from our friends at the Foundation for Ecological Security in Udaipur, our team was able to gather enough data to show the severity of deforestation in the villages and to establish constraints for a new solar cooker; one that would allow solar energy to replace wood as a source of fuel for cooking. Our study revealed that a new design must be:

...durable enough to survive inclement weather conditions.
...able to store solar energy, allowing the cooker to charge during the day and be used by women during normal cooking periods (late at night and early in the morning).
...able to be used inside households.
...easy to operate for the women who will use it.

After our team presented the research at the International Association for the Study of the Commons Biennial Conference in Hyderabad, Professor Udaykumar and his mechanical engineering design project class of thirty students began working to design a solar cook stove that meets the design constraints determined by the research team. At the end of the Spring 2011 semester, one prototype was built. However, there were numerous design ideas that, due to time constraints, were never pursued. It is clear that a lot more help is needed in order to come up with a solution. Because the problem of designing an effective prototype is so complex, it can only be addressed if the numerous design possibilities that exist are pursued. In short, one team is not nearly enough.

That is why Climate Healers is calling on professors, engineering clubs, student organizations, and university groups throughout the country to take up the challenge. The many and invaluable perspectives offered by different groups will allow for an acceptable solution for the villagers. After a suitable prototype is designed, further improvements may be made and the road to regenerating life will have begun. Please feel free to pass this along to any professors or student groups who you believe will want to hear about this challenge.

Designs will be judged by members of the non-profit organization, Engineering for Change, and once the top three designs are determined, the winning teams will have the opportunity to go to the villages in Rajasthan and build their solar cooker.

We would be honored if members of your faculty and student body would take up the challenge and become a part of this collaborative effort. Since its launch last Spring, more and more students are forming teams with the goal of designing a solar cooker for field use. We would be honored if you would join our collective mission. The climate crisis affects all of us, and efforts by world leaders have been unable to address the climate crisis. By employing an approach that involves dedicated groups and individuals, we can reverse the tide of the climate crisis and start undoing the damage.

Thank you very much for your time and for listening to my story...

Sincerely,

Billy Davies"

The response has been very encouraging so far. Not only are several US universities working on the design during this fall semester, but the Indian Institute of Technology in Chennai, India, has posed this cook stove challenge as part of its annual Shaastra festival, in which around 500 colleges and universities participate each year. I'm certain that if there is a viable solution for Rayakbai and her millions of sisters around the world, the Miglets will find it.

9. The Metamorphosis

"Nobody can go back and start a new beginning. But anybody can start today and make a new ending," Maria Robinson..

The Metamorphosis is the transition from the Caterpillar stage to the Butterfly stage for the modern industrial culture. At present, the political left and the political right in America seem to have formed media echo chambers that then revel in pointing out the absurdities of the other side. The political right, while implicitly acknowledging the veracity of computer models by rapidly evacuating from the projected paths of hurricanes, then turns around and pretends that all climate models are bogus and that climate change isn't happening. It has become an article of faith in the right to deny climate change even when entire states are burning up from some of the worst droughts in living memory. The intellectuals on the political left, while excoriating the right for being anti-science, cling to the pretense that infinite economic growth is possible on a finite planet. The trouble is that each side studiously ignores its own pet absurdities, while trumpeting those of the other side.

It is becoming increasingly clear that such a cultural transformation can only occur from the grassroots, beginning with those who are sick of the status quo.

9.1 Miglets and the Corporations

Cultural transformations are primarily led by youth, since adults and senior citizens are usually too habituated to change easily. But the Miglets and the children have been systematically programmed and manipulated to believe that it is the brand of products that they possess and consume that define their self-identity.

Babies are born butterflies. But from the moment they are born, babies become the embryo of the consumer society in our modern industrial culture, with the result that a reverse metamorphosis has been taking place. The children are taught to aim for impossible ideals. They are bombarded with messages on TV, through the internet, in video games, via text messages and twitter feeds, to the tune of over 3000 messages a day for an average American child[1]. The child has now become one of the most influential demographic for the modern corporation to capture.

If a corporation does not spend advertising dollars to ensure brand loyalty with the child, then that corporation will likely not survive. Especially if that corporation is peddling sugary beverages, processed foods, gadgets, bling and products that disconnect the child from reality. Therefore, the corporation has no choice but to plunge into this psycho-social battle to capture the mind of the child, so that the child grows up to be a Pepsi drinker or a Coke drinker even as an adult. In the late seventies, the US Federal Trade Commission (FTC) proposed a ban on all advertising to children[2], but in response, Congress took away the power of the FTC to regulate such advertising to children[3]. And in the eighties, with the deregulatory fervor of the Reagan era and with the advent of cable television, the targeting of children began in earnest.

The Miglets and the children are the victims of this intense marketing blitz. Professional child psychologists and sociologists are employed by corporations to push all the buttons that influence their minds. Child marketing has now become a knowledge discipline with conferences to disseminate information on how to use modern media to target children, how to make age-appropriate graphic violence for the boys and sexualized imagery for the girls.

Writes Prof. Joel Bakan, a law professor at the University of British Columbia[4],

> "Throughout history, societies have struggled with how to deal with children and childhood. In the United States and elsewhere, a broad-based "child saving" movement emerged in the late 19th century to combat widespread child abuse in mines, mills and factories. By the early 20th century, the "century of the child," as a prescient book published in 1909 called it, was in full throttle. Most modern states embraced the

general idea that government had a duty to protect the health, education and welfare of children. Child labor was outlawed, as were the sale and marketing of tobacco, alcohol and pornography to children. Consumer protection laws were enacted to regulate product safety and advertising aimed at children.

By the middle of the century, childhood was a robustly protected legal category. In 1959, the United Nations issued its Declaration of the Rights of the Child. Children were now legal persons; the "best interests of the child" became a touchstone for legal reform.

But the 20th century also witnessed another momentous shift, one that would ultimately threaten the welfare of children: the rise of the for-profit corporation. Lawyers, policy makers and business lobbied successfully for various rights and entitlements traditionally connected, legally, with personhood. New laws recognized corporations as legal — albeit artificial — "persons," granting them many of the same legal rights and privileges as human beings. In an eerie parallel with the child-protective efforts, "the best interests of the corporation" was soon introduced as a legal precept.

A clash between these two newly created legal entities — children and corporations — was, perhaps, inevitable. Century-of-the-child reformers sought to resolve conflicts in favor of children. But over the last 30 years there has been a dramatic reversal: corporate interests now prevail. Deregulation, privatization, weak enforcement of existing regulations and legal and political resistance to new regulations have eroded our ability, as a society, to protect children".

And the Miglets had to grow up in this predatory environment. Corporations appear to be winning this battle. And it isn't even a fair fight as corporations have cornered the political arena through their lobbying and through their control of the financing of elections.

Besides, corporations aren't the only Caterpillars that are preying upon the Miglets. The system itself has been rigged to continue the Caterpillar culture worldwide. In some countries, autocratic rulers usurped absolute power and used that power to amass wealth for themselves and their kin. Even in democratic countries such as India and the US, corruption has become systemic and even legalized. In India, it became next to impossible to get anything done without bribing the government bureaucracy. And in the US, a revolving door between the government bureaucracy and corporations even makes corruption perfectly legal. Governments of the 1%, by the 1% and for the 1% have become firmly ensconced throughout the world to promote the Caterpillar culture[5].

It might appear that it is "Game Over" for the Miglets. That the Caterpillars have won. And the Miglets now have to learn how to survive in a desolate, "Mad Max" world.

But someone forgot to tell the Miglets that they lost. They rose up and drove Hosni Mubarak and his cronies from power in Egypt through a non-violent civil disobedience campaign[6]. They organized and ousted autocrats in Tunisia and Yemen. When asked what they would do if the new rulers also turned out to be autocratic, a Miglet in Egypt said, "We will just come back to Tahrir Square. We did it once and we will do it again".

In India, the Miglets organized and marched in support of Anna Hazare and his hunger strike to root out corruption in the bureaucracy[7]. And the Indian government essentially caved in and acceded to their demands on the Lokpal Bill. Said another Miglet when asked what she would do if the new organization of Lokyuktas also turn out to be corrupt, "We will fast and march again." Likewise, Miglets have also been organizing a five-month long occupation in Madrid, Spain.

But the rebellion of the Miglets has reached a whole new level with the "Occupy Wall Street[8]" and its sister "Occupy Together[9]" actions. By centering the rebellion in Wall Street, the Miglets are confronting the very head of the Caterpillar culture that dominates the planet today. The declaration of the Occupy Wall Street movement reads like a modern version of the Declaration of Independence from 1776, except that corporations have taken on the villainous role that King George played in the original[10]:

"As we gather together in solidarity to express a feeling of mass injustice, we must not lose sight of what brought us together. We write so that all people who feel wronged by the corporate forces of the world can know that we are your allies.

As one people, united, we acknowledge the reality: that the future of the human race requires the cooperation of its members; that our system must protect our rights, and upon corruption of that system, it is up to the individuals to protect their own rights, and those of their neighbors; that a democratic government derives its just power from the people, but corporations do not seek consent to extract wealth from the people and the Earth; and that no true democracy is attainable when the process is determined by economic power. We come to you at a time when corporations, which place profit over people, self-interest over justice, and oppression over equality, run our governments. We have peaceably assembled here, as is our right, to let these facts be known.

They have taken our houses through an illegal foreclosure process, despite not having the original mortgage.

They have taken bailouts from taxpayers with impunity, and continue to give Executives exorbitant bonuses.

They have perpetuated inequality and discrimination in the workplace based on age, the color of one's skin, sex, gender identity and sexual orientation.

They have poisoned the food supply through negligence, and undermined the farming system through monopolization.

They have profited off of the torture, confinement, and cruel treatment of countless animals, and actively hide these practices.

They have continuously sought to strip employees of the right to negotiate for better pay and safer working conditions.

They have held students hostage with tens of thousands of dollars of debt on education, which is itself a human right.

They have consistently outsourced labor and used that outsourcing as leverage to cut workers' healthcare and pay.

They have influenced the courts to achieve the same rights as people, with none of the culpability or responsibility.

They have spent millions of dollars on legal teams that look for ways to get them out of contracts in regards to health insurance.

They have sold our privacy as a commodity.

They have used the military and police force to prevent freedom of the press. They have deliberately declined to recall faulty products endangering lives in pursuit of profit.

They determine economic policy, despite the catastrophic failures their policies have produced and continue to produce.

They have donated large sums of money to politicians, who are responsible for regulating them.

They continue to block alternate forms of energy to keep us dependent on oil.

They continue to block generic forms of medicine that could save people's lives or provide relief in order to protect investments that have already turned a substantial profit.

They have purposely covered up oil spills, accidents, faulty bookkeeping, and inactive ingredients in pursuit of profit.

They purposefully keep people misinformed and fearful through their control of the media.

Carbon Dharma

They have accepted private contracts to murder prisoners even when presented with serious doubts about their guilt.

They have perpetuated colonialism at home and abroad. They have participated in the torture and murder of innocent civilians overseas.

They continue to create weapons of mass destruction in order to receive government contracts. *

To the people of the world,

We, the New York City General Assembly occupying Wall Street in Liberty Square, urge you to assert your power.

Exercise your right to peaceably assemble; occupy public space; create a process to address the problems we face, and generate solutions accessible to everyone.

To all communities that take action and form groups in the spirit of direct democracy, we offer support, documentation, and all of the resources at our disposal.

Join us and make your voices heard!

* These grievances are not all-inclusive".

What is absolutely amazing about these actions is that they are occurring despite the fact that corporations systematically enhanced and manipulated the ego of the Miglets in order to get them to be loyal customers. They are occurring despite the fact that the Miglets were subjected to the quirky theories of the "self-esteem" movement during their school years[11]. The self-esteem movement in education was based on the idea that if children are constantly told that they are smart, they will become capable of learning more. But such unreal feedback only served to increase the narcissistic streak in our children and wound up manipulating and enhancing their ego. This made them acutely aware of their sense of separateness, their acceptance or their rejection from their social cliques, of whether they fit or not.

Says actress Thandie Newton[12],

> "From the age of about five, I knew that I didn't fit. I was the black atheist kid in an all-white Catholic school run by nuns. I was an anomaly. And my self was rooting around for definition trying to plug in. ... We've created entire value systems and a physical reality to support the worth of the self. Look at the industry for self-image and the jobs it creates, the revenue it turns over. We'd be right in assuming that the self is an actual living thing. But it's not; it's a projection, which our clever brains create in order to cheat ourselves from the reality of death.
>
> But there is something that can give the self ultimate and infinite connection -- and that thing is oneness, our essence. The self's struggle for authenticity and definition will never end unless it's connected to its creator -- to you and to me. And that can happen with awareness -- awareness of the reality of oneness and the projection of self-hood. For a start, we can think about all the times when we do lose ourselves. It happens when I dance, when I'm acting. I'm earthed in my essence, and my self is suspended. In those moments, I'm connected to everything -- the ground, the air, the sounds, the energy from the audience. All my senses are alert and alive in much the same way as an infant might feel -- that feeling of oneness".

One way to undergo metamorphosis is to train ourselves to feel that oneness, even if we aren't all inspired dancers like Thandie Newton.

9.2 Awakening through Awareness

In the Vedic view, our sense of oneness with the whole can occur through four main pathways: through our intellect via reasoning (Gnana Yoga), through our emotions, primarily in the form of love and devotion (Bhakti Yoga), through our actions (Karma Yoga) and through meditation and other spiritual practices (Raja Yoga). The word "Yoga" itself, is Sanskrit for "Union," signifying the joining of the self with the whole, the dissolution of separateness and attaining that state of oneness. But in that state of oneness, our thoughts, our emotions, our actions and our entire state of being would be consistent with our awakening so that all these pathways merge into one. Therefore, our journey towards Metamorphosis would be far quicker if we consciously enable all these pathways systematically. Far too often, people enable one pathway to the detriment of the others and fail to reach their goal. For instance, it is common to pursue Bhakti Yoga by singing hymns and bhajans during auspicious times, while our thoughts and actions at other times are uninspired and even detrimental to our goal of awakening.

Awareness or Mindfulness is a tool that we need to lose our sense of separateness, our ego and thus, to awaken. It allows us to reach a state of non-illusion where we begin to see things as they are, not as we are. That is what the Hindus call seeing with the eye of wisdom, the third eye that awakens when the ego is shattered and the Atman is empowered. Hindus normally place a mark on the center of their foreheads that reminds them of that third eye.

In the Bhagavad Gita's model of a human being, the mind, that is the vast "subconscious" mind, of a Caterpillar or Kaurava is running amok, directing the decision-making intellect or the rational mind to fulfill its incessant desires. In a Butterfly or the Pandava, the intellect is in control of the steady mind. Prof. Jonathan Haidt of the University of Virginia has an interesting analogy for the mind and intellect as an Elephant and a rider[13], but since we've already used the Elephant as a metaphor for the Truth, I'll modify it a bit. Imagine a horse and a little child as the rider.

The horse is the mind, while the child rider is the intellect.

In the case of the Caterpillar, the horse is wild, drugged with desires and therefore, bucking around and the child rider is simply hanging on to it for dear life. In the case of the Butterfly, the child is in control of the horse, which is happily trotting along. And awareness is the only tool that we need to change the behavior of that horse.

The book, "Awareness[14]," by Anthony DeMello, is the definitive English book on this subject and both the 4-step processes below are summarized from this book.

Here is a summary of the steps for awakening:

> • **Step 1**: Begin from a position of humility. Acknowledge that you are asleep and want to awaken. Since awakening is a never-ending, infinite process (anyone who claims to be awake is actually asleep!), you are in plenty of company.
> • **Step 2**: Be ready to challenge your existing belief system. It is these ideas that are influencing your life, adding all those illusions and making it the mess that it is.
> • **Step 3**: Be willing to replace these ideas with something unfamiliar, to learn something new.
> • **Step 4**: Become a passive, detached Witness of your thoughts. Don't interfere, don't fix anything, just observe.

It is the Witness in the Upanishadic Witness/Participant duality that can tame the horse. But notice that the Witness can watch his/her thoughts. Then who is this who is aware of the Witness watching the thoughts? And thus, an infinite recursion ensues. This is why the ultimate Witness, the Atman, is unknowable and can only be experienced in the joy of oneness, of integration.

But there are plenty of symptoms that occur when we identify with our ego, the Participant. Just as pain is an indication of something that's not right in our material body, a negative feeling such as anger, frustration, sorrow, etc., is an indication of something that's not right in our mental makeup, that the horse is going crazy again. A negative feeling indicates that we have identified with our ego, the Participant and added some falsehood within us. For the Witness is a compassionate and detached, permanently happy being.

As soon as you get an indication of a negative feeling you can perform the following steps for overcoming them:

- **Step 1**: Identify the negative feeling, become aware of it.
- **Step 2**: Understand that the negative feeling is in you, not in reality. So, stop trying to change reality to fix the negative feeling. No person or event on earth really has the power to disturb or hurt the Witness in you.
- **Step 3**: Never identify with that feeling. It has nothing to do with the Witness and only to do with the Participant and the ego. Let it be, it will pass. Your depressions and thrills have nothing to do with happiness.
- **Step 4**: Change yourself through understanding. When you understand what you didn't agree with in reality or what falsehood you added to reality which caused that negative feeling, then the world becomes right and the feeling will pass.

This doesn't mean that the awakening person will never have a negative feeling, but that the negative feeling will be fleeting if the person practices the above 4-step process as soon as one occurs. Metamorphosis is a process, without end. And as the awareness grows, these negative feelings will occur less and less frequently.

In the beginning, if we are starting off with plenty of negative feelings, then relying on just these four step processes could become tiresome. Therefore, it is best to ground ourselves with a habitual routine for practicing awareness, to calm the horse initially. Here is a variation on a self-compassion exercise from the Stanford Center for Compassion and Altruism Research (CCARE), devised by Prof. Kelly McGonigal[15]. It is a 3-minute breathing and meditation exercise that I've been perfecting over the past year with the help of two dear friends, Dr. Nilima Sabharwal and Dr. Chaya Prasad, to enhance awareness and to dissolve the ego.

The 3-minute segment consists of 10 long breaths for the meditation and 5 short breaths for repose. This segment can be repeated as many times as we like, but it works best if it is repeated at least 10 times during the course of each day. The best way to achieve this is to set aside half an hour each day for meditation at the beginning of the day. In case this is difficult to stick with, an alternate approach is to always take a 3-minute conscious breather before starting every major activity during the day.

Each long breath is a four-step process:

• **Step 1**: Breathe in slowly but steadily over a 4 second period. As you are breathing in, imagine that the energy from the cosmos is coming in through the soles of your feet and filling up your entire body.
• **Step 2**: Hold your breath for 4 seconds. As you are holding your breath, imagine that the energy you took in is gathering up all your negative feelings, all your pain and all your suffering.
• **Step 3**: Breathe out slowly but steadily over a 6 second period. As you are breathing out, imagine that the spent energy is flowing out of your body through the center of your forehead, shattering your ego, taking with it your pain and suffering.
• **Step 4**: Hold your breath for 2 seconds. As you are holding your breath, imagine that your spent energy is dissolving in the infinite energy of the cosmos.

Please don't sweat over the precise seconds for each step. For instance, during each long breath, as a Hindu, I mentally recite the 4 lines of the Gayatri mantra without worrying about the number of seconds. The meanings of the four Sanskrit lines of the Mantra correspond to what is being imagined in each step above during the long breath:

• **Step 1**: (Breathe in) **Om Bhur Bhuva Swaha**. representing Brahma, the Creator. Each intake of breath constitutes the beginning of a new moment of Life and hence, the association with Brahma.
• **Step 2**: (Hold) **Tat Savitr Varenyam**, representing Vishnu, the Sustainer, who is responsible for cleansing away our suffering and sustaining our happiness.
• **Step 3**: (Breathe Out) **Bhargo Devasya Dimahi**, representing Shiva, the Destroyer, who is responsible for destroying our sense of separateness, our ego.

Carbon Dharma

• **Step 4**: (Hold) **DeeyoYona Prachodayat**, representing the supreme Atman, Paramatma or God, the infinite energy of the Cosmos.

The 5 short breaths are the normal 2 second in and 2 second out shallow breath.

I call this conscious breathing and meditation exercise, the **Gayatri Kriya**, because of its association with the Gayatri mantra during the long breaths. For me, the **Gayatri Kriya** is a form of mental hygiene that I must conduct on a daily basis, just as I conduct acts of physical hygiene on a daily basis. Can you imagine going without brushing your teeth or without taking a shower or without excreting for days on end? That would not lead to a healthy body. Likewise, I believe that living without performing some basic rituals of mental hygiene for days on end does not lead to a healthy mind. The half-hour that I spend on the **Gayatri Kriya** every day is a half-hour that has been paying me increasing dividends on a daily basis.

9.3 Call to Action

It would be amazing if we all awakened and the whole world suddenly became enlightened. But that is a truly utopian and perhaps, impractical vision in the near term, especially since we have spent so much energy over the years, as a society, promoting reverse metamorphosis in our children's upbringing. In his Stanford lecture, Prof. Jonathan Haidt of the University of Virginia said that a more practical approach is to constrain the "horse" into a path that it should be taking. That is, if we habituate and constrain ourselves into the actions of a Butterfly, then even the bucking horse may not be so dangerous to the child and to the world.

In the traditional sense, effective activism requires us to physically get out and impede the Caterpillar culture. Humans all over the world have amassed $195 Trillion dollars in wealth with two-thirds of it in the affluent, Western countries[16]. It would be disastrous for Life, in general, to allow all that wealth to be deployed towards the advancement of the pyramid scheme that is the Caterpillar culture.

But impeding the Caterpillar culture is really about playing jiujitsu with it, by turning its strengths into its weaknesses, and utilizing the resources that it provides to advance the cause of the Butterfly.

Acts of grassroots rebellion such as the Tar Sands Action[17] or the Occupy Together action are useful to foster a sense of camaraderie among the disaffected and to persuade the political class to pay attention to their public responsibilities. The Tar Sands Action is a protest to prevent the construction of the Keystone XL Pipeline that would transport the bituminous oil from the Athabasca Tar Sands of Alberta, Canada to refineries in Texas so that the oil can be exported to consumers in Europe and elsewhere. But protesting against the pipeline is somewhat like conducting the "War on Drugs" by fighting the coca-growers of Colombia. This tactic didn't work in the War on Drugs because we didn't reduce the demand for drugs on the consumer side, but only tried to disrupt the supply of drugs. Even when the flow of drugs from Colombia got disrupted, our addicts got hooked on other drugs, even prescription drugs, instead. Therefore, the Tar Sands action must be accompanied by concrete actions on the demand side of the ledger to make it truly effective.

The "Occupy Wall Street" and "Occupy Together" actions are clearly a global continuation of the Miglets in rebellion over the current Caterpillar culture. The main grievance underlying the Occupy Wall Street action is the fact that 1% of Americans have hogged 40% of the financial wealth in America, with Wall Street and corporate executives at the top of the heap. By identifying with the 99% of the people who are not doing so well, the Miglets have captured the imagination of the public. It is truly heart wrenching to see photos of Miglets carrying placards that read, "I have $70K in student loans, $12K in medical bills and I'm 22. Where's my Bailout?" and "I have $40K in student loans and no job. Should I sell drugs or sell my body?" As it is, the Occupy movement is a great start to the Metamorphosis.

When the corporations commodified Nature, we didn't protest too much because they were making gadgets for our entertainment. When the corporations commodified the animals, birds and fishes, we didn't protest too much because they were making meat, milk and eggs for our consumption. When the corporations commodified workers in Third World countries, we didn't protest too much because they were making clothes for our backs. Now that the corporations are commodifying the Miglets in America, the rebellion is on! And the Occupy movement will become the Metamorphosis as it matures and internalizes the understanding that the protesters are really part of the 99.9999% of complex Life that is being commodified by the 0.0001%, the corporations and their executives. It is this understanding that will transform the Occupy movement into a compassionate, positive force for change. For the true Metamorphosis is a joyous undertaking to reconfigure our systems so that the actual division of wealth among the members of our species would be worth arguing about, if at all we even need to, in the years to come.

It is our habits of consumption that fuel the Caterpillar culture. But the objective of our consumption is to create a social world for us and then to define our place in that world. The corporate Caterpillars have deployed a ubiquitous and nearly free communications infrastructure in order to specifically target individual consumers and to promote conspicuous consumption amongst us. It is this communications infrastructure that the Miglets have been using to organize collaboratively and to deploy their grassroots action against the Caterpillars on Wall Street. In essence, the Miglets have already been using jiujitsu techniques against the Caterpillar hierarchy.

What I am suggesting is that we utilize the same jiujitsu approach towards the larger cause of Metamorphosis, to awaken the Butterfly. So far, we have been bombarded with hundreds of things that we can do to "go green". Including things such as burning biofuels in our cars that later turned out to be not so green, after all. But as we did a few of these hundreds of things in our harried daily lives, we discovered that the state of our world was getting worse and worse as hundreds of millions of other human beings were also being simultaneously conscripted into the Caterpillar culture. Therefore, I suggest that we instead take just three practical steps that pack a cohesive punch. These steps are impossible for the Caterpillars to counter and with discipline and perseverance, they will have a significant impact on them. The first two steps can be construed as constraining the "horse," to use the modified metaphor of Prof. Haidt, while the third step is to create a social world in which conscious consumers feel a sense of belonging and acceptance.

The first step is to go vegan, preferably organic, preferably locally grown and preferably with whole, unprocessed foods. I've been consuming mostly organic, and always vegan food for nearly three years and it has been wonderful for my health and well being. But we're also fortunate to live in a Northern California community where farmers' markets are open year round and supermarkets carry organic produce. As a result, we have also minimized our consumption of industrial, processed foods, which Scott Faber, vice president of the Grocery Manufacturer's Association admits[18]: "combine ingredients from hundreds of thousands of suppliers in over 200 countries." The result of such industrial processing could only be what Michael Pollan refers to as "food-like substances[19]," and not real food.

Veganism has seen a tremendous spurt in popularity, recently. The number of vegans on campus has doubled in the United States in the past 4 years. Vegan cuisine has also become increasingly sophisticated. Traditional Asian and especially, Indian cuisines have always been chock full of vegan dishes that are not only healthy, but mouth watering as well. But modern restaurants, such as the Millennium[20] in San Francisco, have created unique dishes that are both artistic and flavorful. Numerous recipe books have been written on vegan cooking and there are tens of thousands of vegan recipes that we can browse for free on the internet[21]. They include recipes on desserts, pastries and other baked items as well. Kelly Peloza, the vegan Cookie Connoisseur[22], is a Miglet who has been revolutionizing vegan baking. And best of all, the vegan Chef Chloe Coscarelli[23], another Miglet, won the Food Network's Cupcake Wars[24], beating out numerous other contestants who were cooking with animal matter.

Chef Chloe is my hero.

Going vegan also means that we purchase only products that don't have any animal content. But that doesn't mean that we have to throw away our existing leather shoes or furniture. I still wear my leather shoes and will do so until they fall off my feet, for two reasons: 1) to remind myself not to be judgmental of others, and 2) to honor the animal that died and gave its skin for my comfort. However, I will never ever buy another animal product in my life.

Which brings me to the second step, which is to counter the impulse buying that is at the heart of the Caterpillar culture. I do this through instituting a "Buy Everything Day" in my life. A few years ago, I received an invitation to join a group that called itself the "Buy Nothing Day[25]" on Facebook. It was a group of people that pledged not to buy anything on the day after Thanksgiving. I was shocked that we had become such compulsive buyers in our society that we needed to remind ourselves not to buy anything on one single day of the year. But it occurred to me that the opposite concept would be extremely powerful to counter the media blitz of advertisers and corporations.

The "Buy Everything Day" is a day that I chose arbitrarily to be the day after my birthday, on which I buy everything I need for the next year. During each year, whenever I think of something that I need, I write it down on a list, but I don't buy it until the next "Buy Everything Day." The only items that I buy on the other days of the year are food items and gasoline, necessities that cannot be stored, but I do make exceptions for downloadable e-books and donations to charity. It turns out that when the Buy Everything Day comes around, I look at my list and cross out most of the items that I had written down except for toothpaste, toothbrush, soap and other grooming necessities. For it is almost impossible for me or, I suspect, for any of us, to spend thousands of dollars on "stuff" on a single day.

The "Buy Everything Day" concept is not new in our family. My parents used to do just that, but they would buy everything on their list once a month and start a new list. I recall the agonizing that they used to go through on certain items as our family budgets were quite tight. But my parents also bought clothes and big ticket items only once a year.

Perhaps, instead of once a year, if the Buy Everything Day occurs once every quarter, this concept would not be so hard to implement for anyone. But even if we think that these first two steps are hard, it is only because they are so counter to our present culture. Change of any kind is hard for a habituated person, but, to paraphrase Henry Ford, "if we think the change is easy or if we think the change is hard, we are usually right." As long as the changed state is conducive to a happy life, then someone is already living it. Therefore, easy and hard are judgements of our mind that either loosen or strengthen the force of our habits, our addictions.

Above all, it is extremely important for us to believe in why we are taking these steps. For if we truly believe in the purpose of the Butterfly, to regenerate Life and undo the damage done by the Caterpillar, then these changes should occur effortlessly[26].

When we change, the world changes with us. If the world doesn't change right away, please don't be discouraged. It will eventually change because of the stone cutter's logic. Perhaps, the hundredth time that we practice the change, the world notices and changes, but it wasn't just our last act that made the world change, but our every past practice mattered as well. But to facilitate the spreading of the change, the third step is to organize frequent meetings, "occupations" as it were, at various locations throughout the world where like-minded people can drop in, participate and strengthen individual resolves. These face-to-face events create the social universe in which conscious consumption of the kind exemplified in the first two steps is celebrated and we can plug in and define our place in that social order. The events should have a common structure that reinforces the four-fold pathway to the awakening of the Butterfly, by exercising the mind, the heart, the intellect and the senses. For example, each session could include:

•meditation to exercise the mind (Raja Yoga),

•discourse to exercise the intellect (Gnana Yoga),

•singing, dancing and sharing in a non-judgmental, interfaith environment to exercise the heart, i.e., our emotions (Bhakti Yoga).

•actions in line with the purpose of the Butterfly in order to exercise our senses (Karma Yoga).

For instance, I attended Occupy SF in San Francisco one day and left impressed with the knowledge and wisdom that was being imparted during their mid-afternoon educational discourses. In that session, the multi-national corporation, Monsanto, was being discussed and any prosecutor worth his/her salt could have developed an indictment of the corporation just by attending. This is why I think that if corporations are expecting the Occupy movement to blow over without a fundamental restructuring of their exploitative, commodifying relationship with the planet and its inhabitants, they are in for a rude awakening.

If implemented with discipline and persistence, these three steps have the potential to truly make a difference. As Margaret Mead said, "Never doubt that a small group of thoughtful, committed citizens can change the world. Indeed, it is the only thing that ever has."

Personally, in any case, I cannot think of anything better to do in our current world situation.

Fig. 7. *"Those who see me in all and all in me are near and dear to me," Bhagavad Gita 6:30.*

Epilogue

"All truth passes through three stages: First, it is ridiculed. Second, it is violently opposed. Third, it is accepted as self-evident," Arthur Schopenhauer.

Dr. Campbell said that I had to sign a document certifying that Midnight hadn't bitten anybody in the past 15 days. Midnight, the gentlest of dogs, had never bitten anybody in more than 4500 days on this Earth.

I suppose that this is a ritual that dog lovers everywhere go through: putting our dogs to sleep. Along with the cancer, Midnight had now developed a degenerative nerve disorder, like Lou Gehrig's disease, and he was having a hard time raising his hips off the ground. It was time to release him from the suffering, from the daily incontinence and from the indignity of having to be lifted off his own excrement in the mornings. For his breed, Midnight was in his 95th year in human terms and had lived long already. I signed the release at the vet's office on Oct. 27, 2011, on my mother's 75th birthday, a fitting day for Midnight to join Amma, who loved all animals as if they were her own children.

It was a little over twelve years ago, in 1999, that our son, Akhil, insisted on a black labrador puppy and we called a breeder in our New Jersey neighborhood and went to see the puppies at his farm. Midnight had such a cute face and expressive eyes and Akhil picked him out as the one to adopt. Midnight was born to a yellow lab mother and a chocolate lab father, but he was pitch black, for Nature is a strange color chemist.

As soon as we got Midnight into our family, Akhil ceased to be the "baby" and Midnight began answering to that moniker as well. For the first year, I used to enjoy taking Midnight out on a daily walk and stop with him as he smelt the story told by every bush on the way. Then he tore his left rear knee ligament while chasing after our children in the front yard and had to undergo surgery. The daily walks were no longer enjoyable for him and he used to stop half way, turn around and pull me back home. He developed arthritis on that knee soon after.

Despite the bad knee and the Irritable Bowel Syndrome (IBS) for which he underwent stomach surgery, Midnight enjoyed playing fetch in the backyard. We adopted a companion for him, Shadow, from the local shelter when Midnight was 3 years old. They had been inseparable since then, except when I used to take Midnight for swimming therapy for his arthritic knee. That dog loved swimming and knew precisely when the last traffic light before the pool was and would start whimpering and getting excited as we waited for the light to turn green. It was with sheer joy that he would wade into the heated pool, looking back to ask me to throw the toy that he could fetch from the pool.

His arthritis got better with the swimming therapy, but it was when we switched him to a vegan diet that he finally stopped limping on that knee. I had read an article about V-dogfood[1], a veterinarian formulated vegan dry dog food, and how it had helped some other dog's arthritis problem and decided to try it out. Soon thereafter, we stopped all the injections that were needed until then to keep Midnight's rear left leg down. Both Midnight and Shadow have been vegan for the past seven years - four years longer than me - and it has been great for their health.

When we moved to California, we rented homes that have pools so that both dogs could swim at home every day, if need be. Though Midnight hadn't availed himself too much of the pool lately, he had stopped limping completely until his demise. It was the veganism that relieved him of the limp in his leg, not the swimming.

Midnight loved our children and was very protective of them. Whenever I yelled at the children, which happened occasionally when they were teenagers, Midnight would bark at me, while wagging his tail furiously. It was his way of telling me to calm down. On happier occasions, the children used to wrestle me to the ground while pretending that I was being rough with them and

Midnight would wade in to rescue them, barking and play-biting me to make me stop.

Though Midnight was gentle with humans and other dogs, he was a menace to machines. He would always growl when we turned on the vacuum cleaner and pushed it towards him. In the beginning, he used to try and catch the vacuum cleaner from behind, to put an end to its mechanical life. Later, after he discovered that we wouldn't let him kill it, he would growl and walk away. But then there was the poor Roomba[2] that lasted precisely one minute in an encounter with Midnight. Roomba was one of those robotic vacuum cleaners that was supposed to clean a room of any shape without supervision. It was a marvel of robotic engineering, invented by a scientist from the Massachusetts Institute of Technology (MIT). But, once Midnight got through with it, the Roomba was upside down on the floor, with all its rubber belts and brushes hanging in tatters. It never vacuumed up a single speck of dust since that fateful encounter.

Midnight and Shadow understood each other and understood us better than we understood them. Midnight would stare me in the face and bark suggestively, waiting for me to "get it," and I would ask him various questions until I hit upon the right one that got him excited. And that's what he was trying to tell me he wanted to do.

It has been three days since I signed the release and said my goodbyes to that handsome dog, but it feels like an eternity. I will miss Midnight tremendously for the rest of my life, but I'm so grateful that he chose to hang out with us on his journey through life.

Thank you, Midnight, for being such a kind, unconditionally loving, gentle dog and for teaching me that animals feel joy and pain, have emotions and experience the universe in their own way, with their unique senses. Thank you, Midnight, for teaching me how to live life to the fullest with joy and abandon, to treat each morsel of food as delicious, to treat each act of play as delightful. Finally, thank you, Midnight, for teaching me about the power of life and death that we, humans, wield over animals, and the dilemma, the "Dharma Sankata," that we face as we wield it.

Rest in peace, my beautiful baby!

If only we could all transfer the compassion we feel for our companion animals towards all other creatures! In the Indian ocean island of Mauritius, the Dodo bird evolved without fear of predation[3]. When it encountered the first humans on the island, it walked up to them innocently and promptly got killed. For the Dodo bird had never encountered a creature that killed for the sake of killing. And pretty soon, the Dodo bird became extinct.

The Dodo bird was not alone. As we now possess technological capabilities that make every creature on the planet as vulnerable to us as the Dodo bird was in the 17th century, we are now consigning tens of thousands of species to the extinction list, each year. But it's time we re-evaluate our tendency to kill first and ask questions later.

I was reminded of the Dodo bird during my technological past, in 2002, the first of my bonus years on this Earth. I was in Haifa, Israel, helping to bring Intel's Gigabit Ethernet networking chips[4] to market and the second intifada was raging throughout Israel. On my way from Tel Aviv's Ben Gurion airport to Haifa, I had seen the smoke from a burned out bus that was the casualty of a suicide bombing. The nervousness over suicide bombings was palpable throughout Israel during those days.

Late one night, after a long and exhausting "debugging" session at the Intel Haifa Lab, I was deep in thought, hurrying along on the boardwalk on the beach towards my hotel and my comfortable bed, carrying my laptop case and wearing my jacket against the cold, when I felt a bright spotlight shining on me. I looked towards the spotlight and could barely make out the silhouette of an Israeli military vehicle on the road that was parallel to the boardwalk. At that moment, I realized the precarious position that I was in. There I was, looking like an Arab, carrying a bulky, black laptop case, and walking towards a posh, seaside resort hotel at a time when suicide bombings were rampant. I was a Dodo bird for staying so late in the Lab during those difficult times. I became terrified that the driver of the vehicle, who was most likely a young Israeli teenager compulsorily drafted into the military, would shoot first and then ask questions later. But that spotlight and the military vehicle followed me for the next five minutes as I walked towards my hotel and entered it, without firing a shot.

Since then, I've always wondered how that soldier decided that it was safe to let me enter the hotel with my bag? Was it the Intel badge that was dangling around my neck, which I had forgotten to take off when I left the office? While I will never know, I still feel tremendously grateful to that soldier for giving me a second lease on Life. And now I savor every moment of that Life.

In Africa, there are tribesmen who trap a baboon by fashioning a hole in an ant hill which is large enough for the baboon to put its hand through, but which is too small for the baboon take its clenched fist out. The tribesman places some treats in the hole in full view of the baboon and retreats behind a tree and waits. The tempted baboon walks up to the hole, reaches into it and gets hold of the treat, but cannot withdraw its fist. And that silly baboon stands there squirming and screaming as the tribesman calmly approaches and captures it. It never occurs to the baboon to let go of the treat and escape[5]. But it remains to be seen whether we, humans, will let go of the animal products that we've gotten used to and escape the approaching tribesman, whether we are truly wiser than the baboon.

For, as a species, we are now in a "Haifa moment," as we have triggered the Sixth Great Mass Extinction event in the Earth's history. As of today, a future paleontologist would deduce a Great Mass extinction event on two-thirds of the ice-free land area of the planet that far surpasses the previous five great Mass extinction events. In all the past five such events on Earth, the dominant, apex predator species didn't survive the event. Here's wishing that as we consciously work towards undoing the damage that we have done so far and towards the betterment of all Life on Earth, we will get through the coming existential constriction without suffering the same fate, without joining the Dodo bird in the fossil record of this era on Earth. That instead, the 21st century will become known as the "Century of the Butterfly," when a documentary about it is made decades from now.

I began this journey four years ago, full of questions, because of my intense love for our two sons, our Miglets. Four years later, I intensely love all the Miglets in the world and indeed, all Life.

Four years ago, we felt financially secure as a family. Four years later, in the aftermath of the economic doldrums in America, we are ex-millionaires and we have officially joined the ranks of the "99%."

Four years ago, I felt as if I was a total failure as a parent and as a human being. Four years later, I've stopped making such judgments and feel truly at peace with myself.

As the saying goes, four years is enough to change anyone.

During these four years, the more I searched for the answers to my questions, the more I realized that they were right under my nose, but I had been too preoccupied to see them. What was happening in my life was a microcosm of what was happening in the world. The Karmic feedback was always there, but I had been too busy with my professional career to "get it." What the late Steve Jobs said in his commencement address at Stanford University is so true[6]: "You can only connect the dots looking backwards."

And I feel truly blessed that the dots in my life connected to lead me on my present journey.

While I don't know how events will unfold into the future, I have complete faith that Life is still trying to reach a balance, to heal the wounds that we have inflicted on it. And that Life is the most powerful force on the planet that can turn things around. Therefore, as long as I'm breathing, I shall believe that things will turn around as the Metamorphosis unfolds. And I intend to use every fiber in my being to help it along.

Carbon Dharma

Our granddaughter, Kimaya, who triggered this literary output, is
now crawling all over the house, but she has still remained a
Butterfly. She has a favorite song, which is from the Hindi movie,
Sant Gyaneshwar[7] from 1964:

Jyoth Se Jyoth Jagathe Chalo (sung by Mukesh):

Awaken the Light within
Let love flow like the river Ganges
If you encounter anyone troubled or sad
Embrace them all and let love flow like the river Ganges

Who's high? Who's low?
He's present in everyone
With the burden of false sentiments
Humans are caught in the web of deceit
Hoist the flag of Dharma and let love flow like the river Ganges

He's ever present in even the smallest particle in the Universe
His light burns eternal in each Being
There is only one Creator and only one Truth
There is only one Supreme Being
While we have this gift of Life, let love flow like the river Ganges

What an appropriate poem for a born Butterfly! But I wish to
conclude this book with another poem by Marilyn Cornelius, a Ph.
D. candidate in the Emmett Interdisciplinary Program in
Environment and Resources (E-IPER) at Stanford University[8]:

The Caterpillar and the Butterfly
a poem by Marilyn Cornelius

inspired
uninhibited
unable to unsee what i
can see

required
exhibited

unable to unbe
what i am
what i must be

a caterpillar eats
the leaf
it lives on
a butterfly
regenerates
the flower
it sips from

we cannot live
while we kill
our source
of life
we cannot breathe
while we allow
our kindred
their strife

which path
will you choose
a caterpillar
of desire
or a butterfly
that lives so lightly
setting the hearts of flowers
on fire...

my path has chosen me...
there is no turning back
from reality

we must embrace a new way
a sea change in culture
not a change
in what we believe
but in what we feel

Carbon Dharma

what we call nature
is not just the ocean
or the forest
nature is you and me
we come from the soil
and to the soil we will return
what will our children say
when they awaken to a world
in which we have left them
to burn?

which path will you choose?
a path to change and healing
or a path to satisfy
whatever you're feeling

we live in one world
and whether our one world
thrives
or even survives
is up to me
and to you

which path will you choose?
the caterpillar eats away at its own life
until one day it transforms
into a butterfly
gracing every flower it touches
for life...

the potential for
being a loving butterfly
dwells within every caterpillar

the potential for flight
is living
it's just a matter of
timing

Carbon Dharma

living for a fleeting time
living so very lightly
and giving all it can
the butterfly makes
now
really count
now is all we ever have...

i want to be that butterfly
so that all may know life
all may live
and thrive

inspired
uninhibited
unable to unsee
what i can see

a world in chaos
hearts craving healing
compassion
as our primary compass
of feeling
and our own Nature
revealing
to us
true liberty...

The Miglets are awakening. And that is such an awesome sight to behold!

In the Mahabharata, Lord Krishna assures Arjuna, "Whatever happened, happened for the best. Whatever is happening, is happening for the best. Whatever will happen, will happen for the best only!" Perhaps there are 7 billion of us now because there is plenty of clean-up work to do around the planet?

Therefore, let's arise, awake and get to work - as Butterflies!

Carbon Dharma

Acknowledgements

"No man is an island, entire of itself; every man is a piece of the continent, a part of the main," John Donne.

I have a tremendous sense of fulfillment at reaching this milestone on my journey, but I would be remiss not to acknowledge all those who cajoled me, supported me and lent me a helping hand whenever I stumbled. I am tremendously grateful to the numerous folks who freely participated in various conversations that shaped my understanding of the human predicament. They contributed through various blogs and social networks, RealClimate, ClimateProgress, DotEarth, The Climate Project and Facebook. Without standing on their shoulders and tapping into their varied perspectives, I couldn't have taken the incremental steps that led to the contents of this book.

I am also immensely grateful to Jaya Row, Paramhansa Yogananda and Purushottam Lal for their wonderful insights into Vedanta and Hinduism, which they so generously dispensed to the world at large through their videos and books. I am thankful to my classmate, Gourish Hosangadi for connecting me with Jaya Row, his sister. It was only after I attended her lecture in Milpitas, CA, that I watched all her videos on Youtube, which helped clear out some of the cobwebs in my mind.

When I look back at the first draft of the book, I realize the tremendous debt of gratitude that I owe my reviewers who helped mold Carbon Dharma into its final form. Billy Davies, Marilyn Cornelius, Juan Jover, Narayan Subramanian, Gani Ganapathi, Joseph Murray, Dan Miller, Mohan Jain, Stuart Scott, Pam Malhotra, Doug Carmichael, Paul Valva, H-S. Udaykumar, Eric Osgood, Jaclyn Richards, Abbey Moffit, Manju Seal, Richard Pauli, Edward Hummel, Umesh Rao, Brian McLaren and Kamal Prasad were truly co-authors of the book as they chiseled away the rough edges and shaped the book into its present form. I am especially grateful to Brian McLaren for writing the Preface, Juan

Jover for highlighting some glaring issues in the early drafts, Manju Seal for suggesting a reordering of chapters that worked better, Paul Valva for being the English teacher and catching many of my grammatical errors, Joe Murray for bringing the Brian Morton article to my attention, Gani Ganapathi, Nilima Sabharwal and Chaya Prasad for convincing me that I have something useful to contribute and lastly, to Marilyn Cornelius for ironing out many of my early, clumsy sentences, for being so incredibly positive about the book from the first draft onwards and for graciously contributing a beautiful poem to conclude the book. Finally, I wish to thank Prof. Thomas Kailath, my Ph. D. advisor at Stanford, for being such a good mentor and for providing me with the backbone to embark on such an endeavor in the first place.

My wife, Jaine, has been a rock in my life. Our love has grown stronger through all our vicissitudes and she has been the best life partner that I could have ever wished for. I'm truly blessed to have her and our family as companions on this wonderful journey called Life.

Sailesh Rao
Danville, CA, Oct. 2011.

Interesting Facts and References

"The woods are lovely, dark and deep, But I have promises to keep, And miles to go before I sleep, And miles to go before I sleep," Robert Frost.

Interesting Facts:

1. While human population increased by a factor of 6 from 1800 to 2000, from 1 billion to 6.1 billion, human consumption increased by a factor of 64 from $400B to $25T.

2. The top 20% of humans were responsible for 83% of world resource consumption while the bottom 20% were responsible for 1.3%, circa the year 2000.

3. The top 1.3 billion people look like 83 billion people from the viewpoint of the bottom 1.3 billion people.

4. The average American home has nearly tripled in size since 1950 and yet there is $20B a year self-storage industry in America to store overflow "stuff" for Americans.

5. The top 500 million people are responsible for half the fossil-fuel based CO2 emissions in the world.

6. Two-thirds of the ice-free land area of the planet has been appropriated for human use, leaving one-third for wildlife.

7. Half the world's forests have been destroyed, mainly in the past 50 years.

8. Three quarters of the marine fisheries have been overfished.

9. 90% of the predatory fish stocks in the ocean have disappeared, mostly eaten in the past 50 years.

10. Species extinctions are occurring at 100-1000 times the background rate.

11. The rate of species extinction is increasing by a factor of 10 every 20 years.

12. The tiger population in the wild has dwindled to 3% of its original strength in the course of the past 100 years.

13. The size of the world economy has more than quadrupled in the past 50 years.

14. Moore's law states that the number of transistors that can be inexpensively placed in a semiconductor would double every 2 years.

15. If Moore's law continued for 300 years, then a semiconductor device will contain more transistors than there are atoms in the solar system.

16. The rate of increase of CO_2 in the atmosphere is now occurring 60 times faster than at any point in the past 650,000 years.

17. The rate of movement of isotherms is occurring an order of magnitude faster than the response rate of ecosystems.

18. The ocean is acidifying an order of magnitude faster than when a mass extinction occurred 55 million years ago.

19. The Earth has experienced a 0.8C (1.4F) rise in temperature over the past 200 years.

20. Humans use nearly one-third of the ice-free land area of the planet just for livestock production.

21. Half the land area of the US is used for livestock production.

22. The world's 1.2 billion heads of cattle consume the equivalent of 8.7 billion human beings.

23. It requires 60 gallons of water to grow one pound of potatoes. It requires 12,000 gallons of water to grow one pound of beef. And, it requires 24,000 gallons of water to grow one pound of grass-fed beef.

24. Humans kill around 60 billion land animals for food each year.

25. Humans consume over 150 million tons of seafood each year.

26. Just over one-fourth of humanity are responsible for most of the consumption of animal foods.

27. it requires 100 Joules of embedded plant-based energy to produce less than 4 Joules worth of animal foods such as eggs, dairy and meat, for human consumption.

28. The average human being is consuming more energy in food than in fuel and shelter combined, when we take into account the embedded plant-based energy input to the animal agriculture systems.

29. The energy required for a vegan diet is one-fourth the energy required for the average world diet and one-seventh the energy required for the affluent diet.

30. The carbon footprint for typical dairy cheese exceeds the carbon footprint for certain meats like pork and chicken.

31. If humans stopped deforestation today, the world's existing forests would sequester half the world's fossil-fuel based CO2 emissions.

32. Half the indigenous cultures that existed 50 years ago are extinct as of today.

33. Of the top 100 economies in the world, 51 are corporations.

34. There are 600 billion large trees in the Amazon rainforest.

35. According to native cultures, the Amazon rainforest has more eyes than leaves.

36. It is estimated that there are 50 billion planets in the Milky way galaxy alone.

37. There are 100 billion to 500 billion galaxies in the observable universe.

38. Health care is a 2.5 trillion dollar a year industry in the US, growing by about 100 billion dollars each year.

39. According to Prof. Colin Campbell, 70-80% of the health care costs in the US can be eliminated if everyone switched to a whole-foods plant-based diet.

40. 2 billion people burn over 1.5 billion tons of firewood annually for cooking alone.

41. The average American child is subjected to around 3000 advertising messages every day.

42. Humans have amassed $195 Trillion in "wealth" with two-thirds of it in Western countries. The true wealth of the planet, the totality and diversity of Life, has been vastly diminished during the course of this "wealth" accumulation.

Interesting Articles:

1. Brian Morton, "Falser Words were never Spoken," New York Times, August 29, 2011.
http://www.nytimes.com/2011/08/30/opinion/falser-words-were-never-spoken.html

2. Joseph Biden, Jr., "China's Rise isn't our Demise," New York Times, September 7, 2011.
http://www.nytimes.com/2011/09/08/opinion/chinas-rise-isnt-our-demise.html

3. Albert Einstein, "Why Socialism?" Monthly Review, May 1949.
http://monthlyreview.org/2009/05/01/why-socialism

4. Robert Goodland and Jeff Anhang, "Livestock and Climate Change: What if the Key Actors in Climate Change are Cows, Pigs and Chicken," Nov. 2009.
http://www.worldwatch.org/files/pdf/Livestock%20and%20Climate%20Change.pdf

5. Jaime Lincoln Kitman, "The Secret History of Lead," Nation, 2000. http://www.thenation.com/article/secret-history-lead

6. Lewis F. Powell Jr., "Confidential Memorandum: Attack of American Free Enterprise System," August 1971. http://reclaimdemocracy.org/corporate_accountability/powell_memo_lewis.html

7. James McWilliams, "Why Eating Meat is Not Personal," Atlanta Journal and Constitution, November 2009. http://www.ajc.com/opinion/heres-my-personal-beef-202065.html

8. Alexandra Horowitz and Ammon Shea, "Think You're Smarter Than Animals? Maybe Not," NY Times Sunday Review, August 20, 2011. http://www.nytimes.com/2011/08/21/opinion/sunday/think-youre-smarter-than-animals-maybe-not.html

9. Joel Bakan, "The Kids Are Not Alright," NY Times The Opinion Pages, August 21, 2011. http://www.nytimes.com/2011/08/22/opinion/corporate-interests-threaten-childrens-welfare.html

10. Joseph Stiglitz, "Of the 1%, by the 1%, for the 1%," Vanity Fair, May 2011. http://www.vanityfair.com/society/features/2011/05/top-one-percent-201105

11. Occupy Wall Street, "Declaration of Occupy Wall Street," Sep. 29, 2011. http://occupywallst.org/forum/first-official-release-from-occupy-wall-street/

12. Dianne Ackerman, "Evolution's Gold Standard," NY Times Opinion Pages, August 8, 2011. http://www.nytimes.com/2011/08/09/opinion/evolutions-gold-standard.html

13. Time Magazine, "Fighting to Save the Earth from Man," Feb. 2, 1970.

http://www.time.com/time/magazine/article/0,9171,878179,00.html
#ixzz1QnqE8fe5

14. Tom Friedman, "The Earth is Full," NY Times Opinion Page,
June 7, 2011.
http://www.nytimes.com/2011/06/08/opinion/08friedman.html

15. Mark Bittman, "Some Animals Are More Equal Than Others,"
NY Times Opinionator Column, March 15, 2011.
http://opinionator.blogs.nytimes.com/2011/03/15/some-animals-are
-more-equal-than-others

16. Paul Krugman, "Another Inside Job," NY Times Opinion Pages,
March 13, 2011.
http://www.nytimes.com/2011/03/14/opinion/14krugman.html

17. Fred Pearce, "The Last Thing Our Hungry World Needs is
More Food," The Daily Mail, Feb. 6, 2011.
http://www.dailymail.co.uk/debate/article-1353810/Beddingtons-pe
rfect-storm-Last-thing-hungry-world-needs-food.html

18. Eduardo Porter, "Why Superstars' Pay Stifles Everyone Else,"
NY Times Business Day, Dec. 25, 2010.
http://www.nytimes.com/2010/12/26/business/26excerpt.html

19. Neil McFarquhar, "African Farmers Displaced as Investors
Move in," NY Times Africa section, Dec. 21, 2010.
http://www.nytimes.com/2010/12/22/world/africa/22mali.html

20. Amy Simmons, "Scientists Fear Mass Extinction As Oceans
Choke," ABC News, Dec. 1, 2010.
http://www.abc.net.au/news/2010-11-30/scientists-fear-mass-extinc
tion-as-oceans-choke/2357322

21. Joel Stein, "The Rise of the Power Vegans," BusinessWeek Nov
4, 2010.
http://www.businessweek.com/magazine/content/10_46/b42031038
62097.htm

Interesting Books:

1. Anthony DeMello, "Awareness: The Perils and Opportunities of Reality," Image Publishers, June 1990.
http://www.amazon.com/Awareness-Opportunities-Reality-Anthony-Mello/dp/0385249373

2. Paul Hawken, "Blessed Unrest: How the Largest Social Movement in History Is Restoring Grace, Justice, and Beauty to the World," Penguin, April 2008.
http://www.amazon.com/Blessed-Unrest-Largest-Movement-Restoring/dp/0143113658

3. Paul R. Ehrlich, "The Population Bomb," Sierra Club and Ballantine Books, 1968.
http://www.amazon.com/Population-Bomb-Paul-R-Ehrlich/dp/1568495870

4. Paramhansa Yogananda, "The Essence of the Bhagavad Gita," Crystal Clarity Publishers, 2006.
http://www.amazon.com/Essence-Bhagavad-Gita-Paramhansa-Remembered/dp/1565892267

5. P. Lal, "The Bhagavad Gita," Lotus Collection Roli Books, 1994.
http://www.rolibooks.com/lotus/lotus-collection/-/the-bhagavad-gita/

6. Edward O. Wilson, "The Creation: An Appeal to Save Life on Earth," W. W. Norton and Company, 2006.
http://www.amazon.com/Creation-Appeal-Save-Life-Earth/dp/0393062171

7. James Gustave Speth, "The Bridge at the Edge of the World: Capitalism, the Environment and Crossing from Crisis to Sustainability," Yale University Press, March 2008.
http://www.amazon.com/qBridge-Edge-World-Environment-Sustainability/dp/0300136110

8. Paul Gilding, "The Great Disruption: Why the Climate Crisis Will Bring On the End of Shopping and the Birth of a New World,"

Bloomsbury Press, March 2011.
http://www.amazon.com/gp/product/1608192237

9. Alan Weisman, "The World Without Us," Thomas Dunne Books, July 2007.
http://www.amazon.com/World-Without-Us-Alan-Weisman/dp/031 2347294

10. Jeremy Rifkin, "The Empathic Civilization: The Race to Global Consciousness in a World in Crisis," Tarcher, Dec. 2009.
http://www.amazon.com/Empathic-Civilization-Global-Consciousn ess-Crisis/dp/1585427659

11. Dan Zarella, "Zarrella's Hierarchy of Contagiousness: The Science, Design, and Engineering of Contagious Ideas," The Domino Project, 2011.
http://www.amazon.com/gp/product/193671924X

12. UN Intergovernmental Panel on Climate Change, "The Fourth Assessment Report," 2007.
http://www.ipcc.ch/publications_and_data/publications_and_data_r eports.shtml

13. Al Gore, "Earth in the Balance: Ecology and the Human Spirit," Plume Publishers, January 1993.
http://www.amazon.com/Earth-Balance-Ecology-Human-Spirit/dp/ 0452269350

14. Mark Lynas, "Six Degrees: Our Future on a Hotter Planet," National Geographic, January 2008.
http://www.amazon.com/Six-Degrees-Future-Hotter-Planet/dp/142 620213X

15. Sylvia A. Earle, "The World is Blue: How Our Fate and the Ocean's Are One," National Geographic, September 2009.
http://www.amazon.com/World-Blue-How-Fate-Oceans/dp/142620 5414

16. UN Report, "Livestock's Long Shadow," 2006.
http://www.fao.org/docrep/010/a0701e/a0701e00.HTM

17. Vernon Heywood, Ed., "Global Biodiversity Assessment," 1995.
http://www.cambridge.org/gb/knowledge/isbn/item5708206/?site_l
ocale=en_GB

18. L. D. Danny Harvey, "Energy and the New Reality 1: Energy Efficiency and the Demand for Energy Services," Routledge, 2010.
http://www.amazon.com/Energy-New-Reality-Efficiency-Services/
dp/1849710724

19. Stefan Wirsenius, "Human Use of Land and Organic Materials: Modeling the Turnover of Biomass in the Global Food System," Chalmers University of Technology and Goteborg University, 2000.
http://www.chalmers.se/ee/EN/personnel/wirsenius-stefan/downloa
dFile/attachedFile_f0/Doctoral_Thesis?nocache=1306401934.49

20. William Leiss, "The Domination of Nature," McGill Queens University Press, 1994.
http://www.amazon.com/Domination-Nature-William-Leiss/dp/077
3511989

21. Francis Bacon, "Novum Organum," 1620.
http://www.constitution.org/bacon/nov_org.htm

22. Jonathan Safran Foer, "Eating Animals," Little, Brown and Company, 2009. http://www.eatinganimals.com/

23. Charles Eisenstein, "The Ascent of Humanity," Panenthea Productions, March 2007. http://www.ascentofhumanity.com/

24. Gro Harlem Brundtland, "Report of the World Commission on Environment and Development: Our Common Future," 1987.
http://www.un-documents.net/wced-ocf.htm

25. James McWilliams, "Just Food: Where Locavores Get It Wrong and How We Can Truly Eat Responsibly," Little, Brown and Company, Aug. 2009.
http://www.amazon.com/Just-Food-Where-Locavores-Responsibly/
dp/031603374X

26. Thomas M. Campbell II and T. Colin Campbell, "The China Study: The Most Comprehensive Study of Nutrition Ever Conducted and the Startling Implications for Diet, Weight Loss and Long-Term Health," BenBella Books, 2004.
http://www.thechinastudy.com/

27. Douglass Carmichael, "GardenWorld Politics: American Values," http://doug.pbworks.com/w/page/18138359/GardenWorld

28. Viktor Frankl, "Man's Search for Meaning: An Introduction to Logotherapy," Beacon Press, 1946.
http://www.amazon.com/Mans-Search-Meaning-Viktor-Frankl/dp/0671023373

29. McKay Jenkins, "What's Gotten Into Us: Staying Healthy in a Toxic World," Random House, 2011.
http://www.amazon.com/dp/1400068037

30. Brian McLaren, "Everything Must Change: Jesus, Global Crises, and a Revolution of Hope," Thomas Nelson, 2007.
http://www.amazon.com/Everything-Must-Change-Global-Revolution/dp/0849901839

31. Jennifer Aaker, Andy Smith with Carlye Adler, "The Dragonfly Effect: Quick, Effective and Powerful Ways to use Social Media to Drive Social Change," Jossey Bass, Sep. 2010.
http://www.amazon.com/gp/product/0470614153

32. Henry David Thoreau, "Walden," Reprinted by Princeton University Press, 1989.
http://www.amazon.com/Walden-Henry-David-Thoreau/dp/0691014647

33. Wayne Pacelle, "The Bond: Our Kinship with Animals and our Call to Defend Them," William Morrow, April 2011.
http://www.amazon.com/dp/0061969788

34. Kate Pickett and Richard Wilkinson, "The Spirit Level: Why Greater Equality Makes Societies Stronger," Bloomsbury Press, Dec. 2009,

http://www.amazon.com/Spirit-Level-Equality-Societies-Stronger/d
p/1608190366

35. Sally G. Bingham, "Love God, Heal Earth: 21 Leading
Religious Voices Speak Out on Our Sacred Duty to Protect the
Environment," St. Lynn's Press, 2009.
http://www.amazon.com/Love-God-Heal-Earth-Environment/dp/09
80028833

36. Richard Heinberg and Daniel Lerch, Editors, "The Post Carbon
Reader: Managing the 21st Century's Sustainability Crises,"
University of California Press, October 2010.
http://www.amazon.com/Post-Carbon-Reader-Managing-Sustainabi
lity/dp/0970950063

37. Dianne Dumanoski, "The End of the Long Summer: Why We
Must Remake Our Civilization to Survive on a Volatile Earth,"
Crown, July 2009.
http://www.amazon.com/End-Long-Summer-Civilization-Volatile/d
p/030739607X

38. Mark Hertsgaard, "HOT: Living Through the Next Fifty Years
on Earth," Houghton Mifflin Harcourt, 2011.
http://www.amazon.com/Hot-Living-Through-Fifty-Years/dp/0618
826122

39. James Hansen, "Storms of My Grandchildren: The Truth About
the Coming Climate Catastrophe and Our Last Chance to Save
Humanity," Bloomsbury, Dec. 2009.
http://www.amazon.com/Storms-My-Grandchildren-Catastrophe-H
umanity/dp/1608192008

40. Thorsten Veblen, "The Theory of the Leisure Class," Penguin
Classics, 1994.
http://www.amazon.com/Theory-Leisure-Penguin-Twentieth-Centu
ry-Classics/dp/0140187952

41. Tyler Cowen, "The Great Stagnation: How America Ate All The
Low-Hanging Fruit of Modern History, Got Sick, and Will
(Eventually) Feel Better," Dutton, Jan. 2011.

http://www.amazon.com/Great-Stagnation-Low-Hanging-Eventuall
y-ebook/dp/B004H0M8QS

42. Raj Patel, "The Value of Nothing: How to Reshape Market
Society and Redefine Democracy," Picador, Jan. 2010.
http://www.amazon.com/Value-Nothing-Reshape-Redefine-Democr
acy/dp/031242924X

43. Bill McKibben, "Eaarth: Making a Life on a Tough New
Planet," Times Books, April 2010.
http://www.amazon.com/Eaarth-Making-Life-Tough-Planet/dp/080
5090568

44. Harriet A. Washington, "Deadly Monopolies: The Shocking
Corporate Takeover of Life Itself--And the Consequences for Your
Health and Our Medical Future," Doubleday, October 2011.
http://www.amazon.com/Deadly-Monopolies-Corporate-Itself-Cons
equences/dp/0385528922

45. Ellen Schultz, "Retirement Heist: How Companies Plunder and
Profit from the Nest Eggs of American Workers," Portfolio, Sep.
2011.
http://www.amazon.com/Retirement-Heist-Companies-Plunder-Am
erican/dp/1591843332

46. Edward O. Wilson "The Diversity of Life." Cambridge, MA:
Harvard University Press, 1992, at
http://www.amazon.com/Diversity-Life-Edward-Wilson/dp/039331
9407

47. David Quammen, "The Song of the Dodo: Island Biogeography
in an Age of Extinction," Schribner, 1997.
http://www.amazon.com/Song-Dodo-Island-Biogeography-Extincti
on/dp/0684827123

Interesting Videos and Documentaries:

1. Jeremy Rifkin, "On Global Issues and Future of the Planet,"
Interview with Een Vandaag,
http://www.youtube.com/watch?v=m9wM-p8wTq4

2. Jaya Row, "Bhagavad Gita Chapters,"
http://www.youtube.com/user/vedantavision#p/c/3F94DFB2BEE78
B0E

3. BBC Documentaries, "The Century of the Self,"
http://www.youtube.com/watch?v=IyPzGUsYyKM

4. Jill Bolte Taylor, "Stroke of Insight,"
http://www.ted.com/talks/jill_bolte_taylor_s_powerful_stroke_of_i
nsight.html

5. Vice President Al Gore, "An Inconvenient Truth,"
http://www.amazon.com/Inconvenient-Truth-Al-Gore/dp/B000ICL
3KG

6. Dan Miller, "A Really Inconvenient Truth,"
http://fora.tv/2009/08/18/A_REALLY_Inconvenient_Truth_Dan_M
iller

7. Stephen Schneider, "Stephen Schneider talks to 52 Climate
Change Skeptics,"
http://www.youtube.com/watch?v=MWgLJrkK8NY

8. Leonardo Di Caprio, "The Eleventh Hour," 2007.
http://topdocumentaryfilms.com/11th-hour/

9. BBC Horizon, "Global Dimming," 2005.
http://topdocumentaryfilms.com/global-dimming/

10. Wade Davis, "Endangered Cultures," 2003.
http://www.ted.com/talks/wade_davis_on_endangered_cultures.htm
l

11. Joaquin Phoenix, Narrator, "Earthlings," 2005.
http://www.earthlings.com/

12. Prof. Antonio Donato Nobre, "Talk at TedXAmazonia," Nov.
2010.
http://tedxtalks.ted.com/video/TEDxAmazonia-Antonio-Donato-No
b

13. T. Colin Campbell, Ph. D., and Caldwell B. Esselstyn Jr., M.D., starring in, "Forks over Knives," Monica Beach Media, 2011. http://www.forksoverknives.com/

14. Don Hardy and Dana Nachmann, producers, "The Human Experiment," KTF Films, 2011. http://www.thehumanexperimentmovie.com

15. Dan Pink, "On the Surprising Science of Motivation," TED talk, 2009. http://www.ted.com/talks/dan_pink_on_motivation.html

16. Sir Ken Robinson, "Schools Kill Creativity," TED talk, 2006, http://www.ted.com/talks/ken_robinson_says_schools_kill_creativity.html

17. Manfred Max-Neef, "Barefoot Economics, Poverty and Why The U.S. is Becoming an Underdeveloping Nation," Democracy Now, Nov. 2010, http://www.democracynow.org/2010/11/26/chilean_economist_manfred_max_neef_on

18. Thandie Newton, "Embracing Otherness, Embracing Myself," TED talk, July 2011. http://www.ted.com/talks/thandie_newton_embracing_otherness_embracing_myself.html

19. Jonathan Haidt, "When Compassion Leads to Sacrilege," Stanford Center for Compassion and Altruism Research, April 2011. http://ccare.stanford.edu/content/jonathan-haidt-when-compassion-leads-sacrilege

20. Kelly McGonigal, "The Power of Self-Compassion," Stanford Cener for Compassion and Altruism Research, Feb.2011. http://kellymcgonigal.com/2011/08/16/the-power-of-self-compassion/

21. Scene from "Animals are Beautiful People," 1974. http://www.youtube.com/watch?v=NxkvY2FZRQg

22. Simon SInek, "How Great Leaders Inspire Action," Sep. 2009. http://www.ted.com/talks/simon_sinek_how_great_leaders_inspire_action.html

23. Jennifer Siebel Newsom et al., "MIss Representation," 2011. http://missrepresentation.org/

24. Vicki Abeles, "Race to Nowhere," A Reel Link Films, 2010. http://www.racetonowhere.com/

25. Adam Curtis, "The Trap: What Happened to Our Dream of Freedom?" BBC Documentaries, http://topdocumentaryfilms.com/the-trap/

26. Richard Attenborough, "Planet Earth: The Complete BBC Series," http://topdocumentaryfilms.com/planet-earth-the-complete-bbc-series/

27. Edward Burtynsky, "Manufactured Landscapes," 2006. http://www.amazon.com/Manufactured-Landscapes-US-Edward-Burtynsky/dp/B000R2GDOS

28. Edward Burtynsky, "Manufactured Landscapes," TED Talk, 2005. http://www.ted.com/talks/edward_burtynsky_on_manufactured_landscapes.html

29. Aaron Scheibner, "A Delicate Balance," Phoenix Films, 2010. http://adelicatebalance.com.au

30. Annie Leonard, "The Story of Stuff." and follow-ons. http://www.storyofstuff.com/

31. Susan Linn, "Consuming Kids," http://topdocumentaryfilms.com/consuming-kids/

32. Steve Jobs, "How to Live Before You Die," Commencement Address at Stanford University, Stanford, CA, June 2005. http://www.ted.com/talks/steve_jobs_how_to_live_before_you_die.html

33. The Living Bridge in Meghalaya, India,
http://www.snotr.com/video/7331/The_Living_Bridge

34. Wangari Mathaai, "I Will Be a Hummingbird,"
http://www.youtube.com/watch?v=IGMW6YWjMxw

35. Sally Erickson and TImothy Scott Bennett, "What a Way to Go: Life at the End of the Empire,"
http://www.youtube.com/watch?v=NNDgXJR7DsY

36. Solar Bottle Lights in the Phillippines,
https://www.youtube.com/watch?v=SBWi3NtND68

37. Bittu Sahgal, "She's Alive... Beautiful... Finite... Hurting... Worth Dying for."
https://www.youtube.com/watch?v=nGeXdv-uPaw

38. Sigourney Weaver, narrator, "ACID TEST: The Global Challenge of Ocean Acidification," NRDC Documentary,
http://vimeo.com/9431503

39. Matt Damon, Narrator, "Inside Job," Sony Classics, 2010.
http://www.sonyclassics.com/insidejob/

40. Toshiro Kanamori, Teacher, "Children Full of Life,"
http://www.youtube.com/watch?v=armP8TfS9Is

41. Krista Tippett, "Reconnecting with Compassion," TED talk, Nov. 2010.
http://www.ted.com/talks/krista_tippett_reconnecting_with_compassion.html

42. Jeremy Jackson, "How We Wrecked the Ocean," TED Talk, 2010. http://www.ted.com/talks/jeremy_jackson.html

43. PBS Documentary, "The Murder of Crows,"
http://video.pbs.org/video/1621910826

Bibliography

Prologue:

[1] Anthony DeMello, "Awareness: The Perils and Opportunities of Reality," Image Publishers, June 1990.http://www.amazon.com/Awareness-Opportunities-Reality-Anthony-Mello/dp/0385249373 . Passage is from the last paragraph in the Chapter on "Self Observation."

[2] Brian Morton, "Falser Words were never Spoken," New York Times, August 29, 2011. http://www.nytimes.com/2011/08/30/opinion/falser-words-were-never-spoken.html

1. The Caterpillar and the Butterfly:

[1] Paul Hawken, "Blessed Unrest: How the Largest Social Movement in History Is Restoring Grace, Justice, and Beauty to the World," Penguin, April 2008. http://www.amazon.com/Blessed-Unrest-Largest-Movement-Restoring/dp/0143113658

[2] More information on SAI Sanctuary can be found at http://saisanctuary.com/

[3] More information on the Foundation for Ecological Security can be found at http://www.fes.org.in/

[4] Information on the Canadian Tar Sands can be found on the National Geographic web site at http://ngm.nationalgeographic.com/2009/03/canadian-oil-sands/kunzig-text

[5] The reclamation of Nature in the abandoned city of Pripyat and Chernobyl is captured in Edward Burtynsky's amazing photographs

in "Manufactured Landscapes," 2006.
http://www.amazon.com/Manufactured-Landscapes-US-Edward-B
urtynsky/dp/B000R2GDOS. More information can be found in
http://abandonedworlds.com/?p=609

[6] Paul R. Ehrlich, "The Population Bomb," Sierra Club and
Ballantine Books, 1968.
http://www.amazon.com/Population-Bomb-Paul-R-Ehrlich/dp/1568
495870. Though Anne Ehrlich was a coauthor on the book and Paul
freely admits it, her name does not appear in the book at the
publisher's insistence.

[7] UN long range population projections can be found in
http://www.un.org/esa/population/publications/longrange2/WorldPo
p2300final.pdf

[8] UN world consumption statistics can be found in the 1996
Human Development Report of the UN Development Project
(UNDP). http://hdr.undp.org/en/reports/global/hdr1996/chapters/

[9] UNDP stats on the disparities in consumption can be found in
the 1996 Human Development Report from [8] and in the 2010
report. http://hdr.undp.org/en/reports/global/hdr2010/ . Dianne
Dumanoski has a good treatment of the population vs. consumption
debate in Chapter 2 of her book, "The End of the Long Summer:
Why We Must Remake Our Civilization to Survive on a Volatile
Earth," Crown, July 2009.
http://www.amazon.com/End-Long-Summer-Civilization-Volatile/d
p/030739607X

[10] Pamela Danziger, "Why People Buy Things They Don't Need:
Understanding and Predicting Consumer Behavior," Kaplan
Publishing, 2004.
http://www.amazon.com/People-Things-They-Dont-Need/dp/07931
86021

[11] The statistics on the self-storage industry can be found in
http://www.selfstorage.org/SSA/Home/AM/ContentManagerNet/C
ontentDisplay.aspx?Section=Home&ContentID=4228 . The

statistics on American home sizes can be found in the NPR report http://www.npr.org/templates/story/story.php?storyId=5525283 .

[12] Joseph Biden, Jr., "China's Rise isn't our Demise," New York Times, September 7, 2011.
http://www.nytimes.com/2011/09/08/opinion/chinas-rise-isnt-our-demise.html

[13] President Herbert Hoover is reported to have made this statement to a group of advertising executives shortly after taking the oath of office in 1929. Here's a NY Times movie review on the "Century of the Self," that reprints President Hoover's quote: http://movies.nytimes.com/2005/08/12/movies/12self.html

[14] Gretchen Daily, Anne and Paul Ehrlich calculated the "Optimum Human Population size" to be 1.5 to 2 billion in http://news.stanford.edu/pr/94/940711Arc4189.html. In the Global Biodiversity Assessment of 1995, it was estimated that the earth can support at most 1 billion people at American levels of consumption. Please see Vernon Heywood, Ed., "Global Biodiversity Assessment," 1995.
http://www.cambridge.org/gb/knowledge/isbn/item5708206/?site_locale=en_GB .

[15] Stephen Pacala, "Equitable Solutions to Greenhouse Warming: On the Distribution of Wealth, Emissions and Responsibility WIthin and Between Nations," Presentation to IIASA, November 2007. http://www.iiasa.ac.at/Admin/PUB/podcast/16pacala.pdf

[16] This particular quote of President George H.W. Bush has been repeated ad nauseum. A good treatment can be found in David R. Loy's paper http://www.interfaithjustpeace.org/pdf/2010/the_nonduality_david_loy.pdf and in Ehrlich and Kennedy, "Millennium Assessment of Human Behavior," Science Magazine, 2005.
http://mahb.stanford.edu/wp-content/uploads/2011/08/4-2005-Ehrlich-Kennedy-MAHB-Science.pdf

[17] Jayanthi Natarajan's quote is taken from the Think Progress article from July 27, 2011.

http://thinkprogress.org/romm/2011/07/27/280330/with-cabinet-res
huffle-is-india-taking-a-new-approach-to-climate-policy/

[18] Jeremy Rifkin, "On Global Issues and Future of the Planet,"
Interview with Een Vandaag,
http://www.youtube.com/watch?v=m9wM-p8wTq4

[19] The quote is taken from a Democracy Now interview of
Michael Moore on September 28, 2011 regarding the Occupy Wall
Street movement.
http://www.democracynow.org/2011/9/28/something_has_started_
michael_moore_on

[20] Kishore Mahbubani was quoted in Tom Friedman's column in
the NY Times on September 7, 2011 at
http://www.nytimes.com/2011/09/07/opinion/friedman-the-whole-tr
uth-and-nothing-but.html . The original article is in the Financial
Times behind a pay wall.

[21] The Gita model of a human being and my interpretation of
Vedanta and the Upanishads draws heavily from Swami Yogananda
Paramhansa's treatment in "The Essence of the Bhagavad Gita,"
Crystal Clarity Publishers, 2006.
http://www.amazon.com/Essence-Bhagavad-Gita-Paramhansa-Rem
embered/dp/1565892267 and from P. Lal's translation of the
Bhagavad Gita in "The Bhagavad Gita," Lotus Collection Roli
Books, 1994.
http://www.rolibooks.com/lotus/lotus-collection/-/the-bhagavad-git
a/. I have also drawn upon Jaya Row's excellent discourses on the
Gita at Bhagavad Gita Chapters,"
http://www.youtube.com/user/vedantavision#p/c/3F94DFB2BEE78
B0E. Unlike most other organized religions, Hinduism does not
recognize a central authority for interpreting Hindu scriptures and
leaves it up to the individual and his/her teacher (Guru) to form
their own opinions. As such, the faith of a Hindu is an expression
of his/her personal freedom. I trust that the reader will take my
interpretation for what it is: an expression of my own limited
understanding of the Hindu scriptures, gleaned mainly through

English translations due to my rudimentary knowledge of the original Sanskrit. Any errors or omissions are strictly my own.

[22] BBC Documentaries, "The Century of the Self," http://www.youtube.com/watch?v=IyPzGUsYyKM .

[23] More information on the India Development Coalition of America can be found at http://www.idc-america.org/ .

[24] This is the famous endosymbiotic theory of Lynn Margulis. Please see, e.g., http://en.wikipedia.org/wiki/Endosymbiotic_theory

[25] Jill Bolte Taylor, "Stroke of Insight," http://www.ted.com/talks/jill_bolte_taylor_s_powerful_stroke_of_i nsight.html

2. Karma:

[1] The statistics on Life Expectancy was taken from an excellent visualization program called GapMinder which crunches on UN data and presents it beautifully. The program can be downloaded from http://www.gapminder.org/ .

[2] The environmental statistics are mainly taken from James Gustave Speth's excellent book, "The Bridge at the Edge of the World: Capitalism, the Environment and Crossing from Crisis to Sustainability," Yale University Press, March 2008. http://www.amazon.com/qBridge-Edge-World-Environment-Sustai nability/dp/0300136110. See also, R. A. Myers and B. Worm, "Rapid Worldwide Depletion of Predatory Fish Communities," Nature 423, 280-283, May 2003, http://www.nature.com/nature/journal/v423/n6937/full/nature01610 .html and UN FAO State of the World's Forests,http://www.fao.org/docrep/011/i0350e/i0350e00.HTM. The statistics on extinction can also be found in Wilson, E. O. 1992. The Diversity of Life. Cambridge, MA: Harvard University Press at http://www.amazon.com/Diversity-Life-Edward-Wilson/dp/039331 9407 .

[3] See, for example, the Hall of Biodiversity at the American Museum of Natural History at http://www.amnh.org/exhibitions/hall_tour/extinct.html .

[4] Edward O. Wilson, "The Creation: An Appeal to Save Life on Earth," W. W. Norton and Company, 2006. http://www.amazon.com/Creation-Appeal-Save-Life-Earth/dp/0393 062171

[5] Charles Dickens wrote this line in "A Tale of Two Cities," Signet Classics, 1997. http://www.amazon.com/Tale-Two-Cities-Signet-Classics/dp/04515 26562

[6] http://en.wikipedia.org/wiki/Thomas_Robert_Malthus

[7] http://en.wikipedia.org/wiki/Haber_process

[8] http://en.wikipedia.org/wiki/Norman_Borlaug

[9] http://en.wikipedia.org/wiki/Dead_zone_(ecology)

[10]http://www.iucn.org/knowledge/news/opinion/?6752/Suddenly-we-find-that-its-all-gone

[11] James Gustave Speth, "The Bridge at the Edge of the World: Capitalism, the Environment and Crossing from Crisis to Sustainability," Yale University Press, March 2008. http://www.amazon.com/qBridge-Edge-World-Environment-Sustai nability/dp/0300136110

[12] Paul Gilding, "The Great Disruption: Why the Climate Crisis Will Bring On the End of Shopping and the Birth of a New World," Bloomsbury Press, March 2011. http://www.amazon.com/gp/product/1608192237

[13]http://en.wikipedia.org/wiki/United_States_Declaration_of_Inde pendence

[14]http://en.wikiquote.org/wiki/Philip_K._Dick#.22How_To_Build_A_Universe_That_Doesn.27t_Fall_Apart_Two_Days_Later.22_.2_81978.29

[15] Thomas Day's quote can be found in Armitage, "The Declaration of Independence: A Global History," Harvard University Press, 2007. http://www.amazon.com/Declaration-Independence-Global-History/dp/0674022823 . The quote can be found in http://en.wikipedia.org/wiki/United_States_Declaration_of_Independence as well.

[16] Stephen Douglas's quote can be found in Pauline Maier, "American Scripture: Making the Declaration of Independence," Knopf, 1997. http://www.amazon.com/American-Scripture-Making-Declaration-Independence/dp/0679779086 . The quote can be found inhttp://en.wikipedia.org/wiki/United_States_Declaration_of_Independence as well.

[17] http://en.wikiquote.org/wiki/Martin_Luther_King,_Jr.

[18] Bhagavad Gita 2:47 in P. Lal, "The Bhagavad Gita," Lotus Collection Roli Books, 1994.http://www.rolibooks.com/lotus/lotus-collection/-/the-bhagavad-gita/

3. Dharma:

[1] Alan Weisman, "The World Without Us," Thomas Dunne Books, July 2007. http://www.amazon.com/World-Without-Us-Alan-Weisman/dp/0312347294

[2] The Cosmic Fig Tree Story is drawn from P. Lal's excellent exposition in the Introduction of "The Bhagavad Gita," Lotus Collection Roli Books, 1994. http://www.rolibooks.com/lotus/lotus-collection/-/the-bhagavad-gita/

[3] Jeremy Rifkin, "The Empathic Civilization: The Race to Global Consciousness in a World in Crisis," Tarcher, Dec. 2009.
http://www.amazon.com/Empathic-Civilization-Global-Consciousness-Crisis/dp/1585427659

[4] Albert Einstein, "Why Socialism?" Monthly Review, May 1949.
http://monthlyreview.org/2009/05/01/why-socialism

[5] Anthony DeMello, "Awareness: The Perils and Opportunities of Reality," Image Publishers, June 1990.
http://www.amazon.com/Awareness-Opportunities-Reality-Anthony-Mello/dp/0385249373

[6] Dan Zarella, "Zarrella's Hierarchy of Contagiousness: The Science, Design, and Engineering of Contagious Ideas," The Domino Project, 2011.
http://www.amazon.com/gp/product/193671924X

[7] http://en.wikipedia.org/wiki/Moore's_law

4. The Really Inconvenient Truth:

[1] http://www.linktv.org/

[2] http://en.wikipedia.org/wiki/The_Climate_Project

[3] Vice President Al Gore, "An Inconvenient Truth,"
http://www.amazon.com/Inconvenient-Truth-Al-Gore/dp/B000ICL3KGand Vice President Al Gore, "An Inconvenient Truth,"
http://www.amazon.com/Inconvenient-Truth-Al-Gore/dp/B000ICL3KG

[4] http://www.ipcc.ch/

[5] Al Gore, "Earth in the Balance: Ecology and the Human Spirit," Plume Publishers, January 1993.
http://www.amazon.com/Earth-Balance-Ecology-Human-Spirit/dp/0452269350

[6] Dan Miller, "A Really Inconvenient Truth," http://fora.tv/2009/08/18/A_REALLY_Inconvenient_Truth_Dan_M iller

[7] http://en.wikipedia.org/wiki/Deepwater_Horizon_oil_spill

[8]http://en.wikipedia.org/wiki/Fukushima_Daiichi_nuclear_disaster

[9] http://climaterealityproject.org/

[10] An excellent distillation of the University of Washington's PIOMAS data can be found in http://neven1.typepad.com/blog/volume-and-concentration/ .

[11]http://thinkprogress.org/politics/2010/11/04/128346/rove-climate -is-gone/

[12] Information on the Canadian Tar Sands can be found on the National Geographic web site at http://ngm.nationalgeographic.com/2009/03/canadian-oil-sands/kun zig-text

[13] Stephen Schneider, "Stephen Schneider talks to 52 Climate Change Skeptics," http://www.youtube.com/watch?v=MWgLJrkK8NY

[14]http://en.wikipedia.org/wiki/Carbon_dioxide_in_Earth's_atmosp here

[15] Hansen et al., "Global Temperature Change," Proceedings of the National Academy of Sciences, July 2006. http://www.pnas.org/content/103/39/14288.full

[16] A. Ridgewell, D. Schmidt, "Past constraints on the vulnerability of marine calcifiers to massive carbon dioxide release," Nature Geosciences, 2010. http://www.nature.com/ngeo/journal/v3/n3/abs/ngeo755.html

[17] Hans Joaquin Schellnhuber interview in Australia. http://www.abc.net.au/lateline/content/2011/s3268037.htm

[18] Compiled from http://en.wikipedia.org/wiki/Thermoregulation

[19] Compiled from Mark Lynas, "Six Degrees: Our Future on a Hotter Planet," National Geographic, January 2008. http://www.amazon.com/Six-Degrees-Future-Hotter-Planet/dp/142620213X

[20] Normal body temperatures of various animals can be found in http://www.fao.org/docrep/t0690e/t0690e04.htm

[21] Sokolov et al., "Probabilistic Forecast for 21st Century Climate Based on Uncertainties in Emissions (without Policy) and Climate Parameters," MIT Report, 2009. http://globalchange.mit.edu/pubs/abstract.php?publication_id=990

[22] Richard Betts et al., "4C Warming: Regional Patterns and Timing," 4C+ Conference, Oxford University, 2009. http://www.eci.ox.ac.uk/4degrees/ppt/1-2betts.pdf

[23] Sylvia A. Earle, "The World is Blue: How Our Fate and the Ocean's Are One," National Geographic, September 2009. http://www.amazon.com/World-Blue-How-Fate-Oceans/dp/1426205414

[24] Richard L. Wallace, "Market Cows: A Potential Profit Center," Illini Dairy Net Papers, 2002. http://www.livestocktrail.uiuc.edu/dairynet/paperdisplay.cfm?contentid=354

[25] "Birds Could Signal Mass Extinction," Oxford University study, 2010. http://www.physorg.com/news205483725.html

[26] UN Report, "Livestock's Long Shadow," 2006. http://www.fao.org/docrep/010/a0701e/a0701e00.HTM

[27] D. Pimentel and M. Pimentel, "Sustainability of Meat-Based and Plant-Based Diets and the Environment," American Journal of Clinical Nutrition, Sep. 2003. http://www.ajcn.org/content/78/3/660S.full

[28] Vernon Heywood, Ed., "Global Biodiversity Assessment," 1995.
http://www.cambridge.org/gb/knowledge/isbn/item5708206/?site_l
ocale=en_GB

[29] Al Gore, "Interview with Larry King," 2009.
http://www.youtube.com/watch?v=_n8FXcJ1mtM

[30] James Hansen, "Storms of My Grandchildren: The Truth About
the Coming Climate Catastrophe and Our Last Chance to Save
Humanity," Bloomsbury, Dec. 2009.
http://www.amazon.com/Storms-My-Grandchildren-Catastrophe-H
umanity/dp/1608192008

[31] C. Milesi et al., "Climate Variability, Vegetation Productivity
and People at Risk," Global and Planetary Change, 2005.
http://secure.ntsg.umt.edu/publications/2005/MHRN05/milesi_et_a
l_GPC_2005.pdf

[32] Robert Goodland and Jeff Anhang, "Livestock and Climate
Change: What if the Key Actors in Climate Change are Cows, Pigs
and Chicken," Nov. 2009.
http://www.worldwatch.org/files/pdf/Livestock%20and%20Climate
%20Change.pdf

[33] See Chapter 7 in L. D. Danny Harvey, "Energy and the New
Reality 1: Energy Efficiency and the Demand for Energy Services,"
Routledge, 2010.
http://www.amazon.com/Energy-New-Reality-Efficiency-Services/
dp/1849710724

[34] Stefan Wirsenius, "Human Use of Land and Organic Materials:
Modeling the Turnover of Biomass in the Global Food System,"
Chalmers University of Technology and Goteborg University, 2000.
http://www.chalmers.se/ee/EN/personnel/wirsenius-stefan/downloa
dFile/attachedFile_f0/Doctoral_Thesis?nocache=1306401934.49

[35] The quote is from the book by Richard Dawkins, "A Devil's
Chaplain: Reflections on Hope, Lies, Science, and Love,"
Houghton, Mifflin and Harcourt, 2003.

http://www.amazon.com/Devils-Chaplain-Reflections-Hope-Scienc
e/dp/0618335404

[36] Edward O. WIlson interview at UNESCO, "The Loss of
Biodiversity is a Tragedy,"
http://www.unesco.org/new/en/media-services/single-view/news/ed
ward_o_wilson_the_loss_of_biodiversity_is_a_tragedy/

[37] Mark Bittman, "Hooked on Meat," NY Times Opinionator, May
31, 2011.
http://opinionator.blogs.nytimes.com/2011/05/31/meat-why-bother/

[38] Jeremy Rifkin, "On Global Issues and Future of the Planet,"
Interview with Een Vandaag,
http://www.youtube.com/watch?v=m9wM-p8wTq4

[39] BBC Horizon, "Global Dimming," 2005.
http://topdocumentaryfilms.com/global-dimming/

[40] Quote is taken from the Economist Magazine article, "Beating a
Retreat: Arctic Sea Ice is Melting Faster than Climate Models
Predict. Why?" Sep. 24,
2011. http://www.economist.com/node/21530079

[41] Quote is taken from Marlowe Hood, "Forests Soak up Third of
Fossil Fuel Emissions," Cosmos Magazine, July 15, 2011.
http://www.cosmosmagazine.com/news/4524/forests-soak-third-fos
sil-fuel-emissions

5. The Kurukshetra of Our TImes:

[1] http://en.wikipedia.org/wiki/Hindu_calendar

[2] http://en.wikipedia.org/wiki/Indian_Institutes_of_Technology

[3] This section on the symbolism of Idols is drawn from Swami
Yogananda Paramhansa's treatment in "The Essence of the
Bhagavad Gita," Crystal Clarity Publishers, 2006.
http://www.amazon.com/Essence-Bhagavad-Gita-Paramhansa-Rem
embered/dp/1565892267 and through osmosis from various other
sources. The reader is welcome to interpret the idols in any way

they like, but it is important to acknowledge that the depiction of fantastic figures with unusual limb and facial arrangements is not meant to be taken literally.

[4]http://en.wikisource.org/wiki/The_poems_of_John_Godfrey_Saxe/The_Blind_Men_and_the_Elephant

[5] This section draws heavily from the discourses of Jaya Row at "Bhagavad Gita Chapters," http://www.youtube.com/user/vedantavision#p/c/3F94DFB2BEE78B0E but, once again, the synthesis and any resulting errors and omissions are my own.

6. The Caterpillar Culture:

[1] Wade Davis, "Endangered Cultures," 2003. http://www.ted.com/talks/wade_davis_on_endangered_cultures.html

[2] Laura Del Col, "The Life of the Industrial Worker in NIneteenth Century England," West Virginia University, http://www.victorianweb.org/history/workers1.html

[3] Ranking as of 2000 taken from Sarah Anderson and John Cavanaugh, "Report on the Top 200 Corporations," from Institute for Policy Studies, 2000. http://www.corporations.org/system/top100.html

[4] http://en.wikipedia.org/wiki/Late-2000s_financial_crisis

[5] Reiner Grundmann, "The Protection of the Ozone Layer," UN Vision Project on Global Public Policy Networks, 2000. http://www.gppi.net/fileadmin/gppi/Grundmann_Ozone_Layer.pdf

[6] Jaime Lincoln Kitman, "The Secret History of Lead," Nation, 2000. http://www.thenation.com/article/secret-history-lead

[7] Attributed to Adam Smith by Charles Handy, founder of the London Business School. Please see, e.g., http://dotearth.blogs.nytimes.com/2008/02/06/the-endless-pursuit-of-unnecessary-things/

[8] Please see section on "Abusing Smith's statement of an invisible hand," at http://en.wikipedia.org/wiki/Invisible_hand

[9] William Leiss, "The Domination of Nature," McGill Queens University Press, 1994. http://www.amazon.com/Domination-Nature-William-Leiss/dp/0773511989

[10] Francis Bacon, "Novum Organum," 1620. http://www.constitution.org/bacon/nov_org.htm

[11] Jonathan Safran Foer, "Eating Animals," Little, Brown and Company, 2009. http://www.eatinganimals.com/

[12] The Humane Society of the United States has compiled a video on the Canadian Seal Hunt at http://www.youtube.com/watch?v=WeYPM8ncsR8

[13] A video on the Calderon Dolphin slaughter in the Faroe Islands of Denmark can be found in http://www.protecttheocean.com/denmark-continues-dolphin-slaughter-warning-graphic-images/

[14] Statistics taken from Aaron Scheibner, "A Delicate Balance," Phoenix Films, 2010. http://adelicatebalance.com.au . The interested reader can compile their own statistics from the Food and Agricultural Organization (FAO) web site: http://faostat.fao.org/site/569/default.aspx#ancor . Please select "World + (Total)" under Geographic groupings as country, select 2009 as the year, "Chicken Meat" as the item and "Producing Animals Slaughtered" as the element and you will find over 52 billion slaughtered in that year. The counting of sea creatures is trickier, but the estimate of 90 billion is likely to be very low. One estimate is that 59 billion land and sea animals are killed in the US alone. Please see, e.g., http://freefromharm.org/farm-animal-welfare/59-billion-land-and-sea-animals-killed-for-food-in-the-us-in-2009/

[15] Please see the PETA page on the mistreatment of circus animals at
http://www.peta.org/issues/animals-in-entertainment/circuses.aspx

[16] Joaquin Phoenix, Narrator, "Earthlings," 2005.
http://www.earthlings.com/

[17] http://en.wikipedia.org/wiki/Traditional_Chinese_medicine

[18] Asia One News, "Mother Bear Kills Cub and then Itself," Aug 5, 2011.
http://www.asiaone.com/News/Latest%2BNews/Asia/Story/A1Story20110805-292947.html

[19] The estimate of Net Primary Production captured for human use is from Milesi et al., "Climate Variability, Vegetation Productivity and People at Risk," Global and Planetary Change, 2005.
http://secure.ntsg.umt.edu/publications/2005/MHRN05/milesi_et_al_GPC_2005.pdf

[20] Please see, e.g., answer to Question 8 on WGBH Boston's FAQ page at http://www.pbs.org/wgbh/evolution/library/faq/cat01.html .

[21] Richard White, ""It is your Misfortune and None of my Own": A New History of the American West," University of Oklahoma Press, 1993.
http://www.amazon.com/Its-Your-Misfortune-None-Own/dp/0806125675

[22] Here's a contemporary newspaper article on the CITES decision:
http://www.guardian.co.uk/environment/2010/mar/18/bluefin-tuna-un-cites

[23] Here's a video on the Bluefin Tuna auction:
http://www.youtube.com/watch?v=JHYm4zFZ8vc

[24] Please see, e.g., Jan Oosthoek's article on the Colonial Origins of Scientific Forestry in
http://www.eh-resources.org/colonial_forestry.html

[25] Here's the US Supreme Court web page concerning the Patent clause:
http://supreme.justia.com/constitution/article-1/40-copyrights-and-patents.html

[26] See the section on Trust Busting at
http://en.wikipedia.org/wiki/Presidency_of_Theodore_Roosevelt

[27] http://saregamapashow.com/

[28] http://en.wikipedia.org/wiki/Pyramid_scheme

[29] Joe Romm who runs the popular ClimateProgress blog wrote an article on the Global Economic Ponzi Scheme or Pyramid Scheme.
http://thinkprogress.org/romm/2009/03/08/203784/ponzi-scheme-madoff-friedman-natural-capital-renewable-resources/

[30] Charles Eisenstein, "The Ascent of Humanity," Panenthea Productions, March 2007. http://www.ascentofhumanity.com/

[31] Here's the Time Magazine article from Sep. 21, 2001 on President Bush's speech, referring to the "shopping,"
http://www.time.com/time/nation/article/0,8599,175757,00.html

[32] Excerpted from
http://www.jfklibrary.org/Research/Ready-Reference/RFK-Speeches/Remarks-of-Robert-F-Kennedy-at-the-University-of-Kansas-March-18-1968.aspx

[33]http://en.wikipedia.org/wiki/Assassination_of_Martin_Luther_King,_Jr.

[34] http://en.wikipedia.org/wiki/May_1968_in_France

[35]
http://en.wikipedia.org/wiki/Assassination_of_Robert_F._Kennedy

[36] http://en.wikipedia.org/wiki/French_legislative_election,_1968

[37] The text of the Powell Memo can be found in
http://reclaimdemocracy.org/corporate_accountability/powell_mem
o_lewis.html

[38] Here's a short list of Bell Labs top inventions:
http://www.usatoday.com/tech/news/2006-12-01-bell-research_x.ht
m

[39] The text of the Senate resolution and the vote tally can be found
in http://www.nationalcenter.org/KyotoSenate.html

7. The Butterfly Culture:

[1] The Planetary Society pinpoints the modern Search for
Extraterrestrial Intelligence to the paper published in 1959 by
Cocconi and Morrison. Please see, e.g.,
http://planetary.org/explore/topics/seti/seti_history_02.html

[2] Here are the web resources for the SETI at Home project.
http://setiathome.berkeley.edu/

[3] M. Mayer and D. Queloz, "A Jupitcr-mass Companion to a
Solar-type Star," Nature 378, 1995.
http://www.nature.com/nature/journal/v378/n6555/abs/378355a0.ht
ml

[4] The estimate of the number of planets in the Milky Way can be
found in http://en.wikipedia.org/wiki/Milky_Way . Estimates of the
number of galaxies in the Universe varies. Edwin Hubble's estimate
was 200 billion galaxies.

[5] International Business Times, "Planet Made of Diamonds
DIscovered 4,000 Light Years Away," August 26, 2011.
http://www.ibtimes.com/articles/204319/20110826/planet-made-of-
diamonds-discovered-diamond-planet-university-of-manchester-ear
th-serpens-oxygen-carb.htm

[6] Gro Harlem Brundtland, "Report of the World Commission on
Environment and Development: Our Common Future," 1987.
http://www.un-documents.net/wced-ocf.htm

[7] Quote found in Chris Lydon, "Real India: A Land Soon WIthout Tigers and Maybe Orchids," August 3, 2010.
http://www.globalconversation.org/2010/08/03/real-india-land-soon-without-tigers-and-maybe-orchids

[8] Prof. Antonio Donato Nobre, "Talk at TedXAmazonia," Nov. 2010.
http://tedxtalks.ted.com/video/TEDxAmazonia-Antonio-Donato-Nob

[9] Statistics taken from Chapter 7 of L. D. Danny Harvey, "Energy and the New Reality 1: Energy Efficiency and the Demand for Energy Services," Routledge, 2010.
http://www.amazon.com/Energy-New-Reality-Efficiency-Services/dp/1849710724

[10] James McWilliams, "Why Eating Meat is Not Personal," Atlanta Journal and Constitution, November 2009.
http://www.ajc.com/opinion/heres-my-personal-beef-202065.html

[11] From Ralph Waldo Emerson, "The Conduct of Life," 1860.
http://www.amazon.com/Conduct-Life-Ralph-Waldo-Emerson/dp/0761834117

[12]http://www.ucsusa.org/assets/images/si/science-idol-2011/web-KLOSSNER-UCS2012calendarCOLOR.jpg

[13] T. Colin Campbell, Ph. D., and Caldwell B. Esselstyn Jr., M.D., starring in, "Forks over Knives," Monica Beach Media, 2011.
http://www.forksoverknives.com/

[14] John Fritze, "Medical Expense Have Very Steep Rate of Growth," Feb 4, 2010, USA Today.
http://www.usatoday.com/news/health/2010-02-04-health-care-costs_N.htm

[15] Quote taken from Anthony DeMello, "Awareness: The Perils and Opportunities of Reality," Image Publishers, June 1990.
http://www.amazon.com/Awareness-Opportunities-Reality-Anthony-Mello/dp/0385249373

[16] The Meatless Monday campaign is organized out of
http://www.meatlessmonday.com/

[17] The Environmental Working Group has a lifecycle assessment
of the carbon intensity of various foods at
http://breakingnews.ewg.org/meateatersguide/a-meat-eaters-guide-t
o-climate-change-health-what-you-eat-matters/climate-and-environ
mental-impacts/

[18] See, e.g., Dr. Jane Parish, "Cow Culling Decisions," Extension
Beef Cattle Specialist, Mississippi State University, 2006.
http://www.thebeefsite.com/articles/664/cow-culling-decisions

[19] The water footprint of grass-fed beef is found in D. Pimentel
and M. Pimentel, "Sustainability of Meat-Based and Plant-Based
Diets and the Environment," American Journal of Clinical
Nutrition, Sep. 2003. http://www.ajcn.org/content/78/3/660S.full

[20] This is a process for converting biomass into fertile soil using
worms. Please see, e.g., http://en.wikipedia.org/wiki/Vermicompost

[21] Such burning was popular during the time of the Emperor Nero.
Please see, e.g.,
http://www.roman-colosseum.info/colosseum/roman-executions-at-
the-colosseum.htm

[22] The sordid history of lynching in the United States can be found
in http://en.wikipedia.org/wiki/Lynching_in_the_United_States

[23] For more on Cannibalism, please see, e.g.,
http://en.wikipedia.org/wiki/Cannibalism

[24] The Occupy Wall Street movement resides on the web at
http://occupywallst.org/

[25] Don Hardy and Dana Nachmann, producers, "The Human
Experiment," KTF Films, 2011.
http://www.thehumanexperimentmovie.com

[26] The Transition Town movement resides on the web at
http://www.transitionnetwork.org/

[27] Here's a video of Amory Lovins describing how he grows bananas in Colorado:
http://www.youtube.com/watch?v=mJbVLyst4Ok

[28] Douglass Carmichael, "GardenWorld Politics: American Values," http://doug.pbworks.com/w/page/18138359/GardenWorld

[29] A History of Bottled Water in the US can be found in http://water.columbia.edu/?id=learn_more&navid=bottled_water. Please see also http://storyofstuff.org/bottledwater/

[30] Solvie Karlstrom and Christine Dell'Amore, "Why Tap Water is Better than Bottled Water," National Geographic Daily News, March 10, 2010.
http://news.nationalgeographic.com/news/2010/03/100310/why-tap-water-is-better/

[31] https://www.linux.com/

[32] http://www.wikipedia.org/

[33] http://www.charityfocus.org/new/

[34] http://www.khanacademy.org/

[35] Dan Pink, "On the Surprising Science of Motivation," TED talk, 2009. http://www.ted.com/talks/dan_pink_on_motivation.html

[36] Sir Ken Robinson, "Schools Kill Creativity," TED talk, 2006, http://www.ted.com/talks/ken_robinson_says_schools_kill_creativity.html

[37] Manfred Max-Neef, "Barefoot Economics, Poverty and Why The U.S. is Becoming an Underdeveloping Nation," Democracy Now, Nov. 2010,
http://www.democracynow.org/2010/11/26/chilean_economist_manfred_max_neef_on

[38] Viktor Frankl, "Man's Search for Meaning: An Introduction to Logotherapy," Beacon Press, 1946.

http://www.amazon.com/Mans-Search-Meaning-Viktor-Frankl/dp/0
671023373

[39] Alexandra Horowitz and Ammon Shea, "Think You're Smarter
Than Animals? Maybe Not," NY Times Sunday Review, August 20,
2011.
http://www.nytimes.com/2011/08/21/opinion/sunday/think-youre-s
marter-than-animals-maybe-not.html

[40] Joel Achenbach, "Zoo Mystery: How did Apes and Birds know
Quake was Coming," WashingtonPost, August 24, 2011.
http://www.washingtonpost.com/national/health-science/zoo-myste
ry-how-did-apes-and-birds-know-quake-was-coming/2011/08/24/gI
QAZrXQcJ_story.html

[41] Matt Woolsey, "Inside the World's First Billion Dollar Home,"
Forbes Magazine, April 30, 2008.
http://www.forbes.com/2008/04/30/home-india-billion-forbeslife-c
x_mw_0430realestate.html

8. Climate Healers:

[1] Gothekar Pada, "Putting the Smallest FIrst: Why India Makes a
Poor Fist of Feeding the Young and How it Can do Better," The
Economist magazine, Sep. 23, 2010.
http://www.economist.com/node/17090948

[2] http://www.climatehealers.org

[3] http://www.thelightingproject.org/

[4] The quote can be found in
http://en.wikipedia.org/wiki/Thomas_Edison

[5] http://www.bogolight.com/

[6] http://www.fes.org.in/

[7]http://en.wikipedia.org/wiki/Kumbhalgarh_Wildlife_Sanctuary

[8]http://en.wikinews.org/wiki/Fresh_floods_in_India_claims_sixtee
n_lives

[9] This estimate is based on the FAO projection that half the wood
consumed annually is burnt as fuel worldwide.
http://www.fao.org/docrep/004/y3557e/y3557e10.htm .

[10] From the Climate Healers Project Report from August 2010,
which can be downloaded from
https://www.engineeringforchange.org/workspace/view/22/1#tabs=/
workspace/files/22/1

[11] http://nrega.nic.in/netnrega/home.aspx

[12] http://www.uiowa.edu/~geog/india/

[13] http://abbeymoffitt.wordpress.com/

[14]https://www.engineeringforchange.org/news/2011/03/01/challen
ge_make_a_solar_cookstove_that_works_at_night.html

9. The Metamorphosis:

[1] George Comstock and Erica Scharrer, "The Media and the
American Child," Academic Press, 2007.
http://www.amazon.com/Media-American-Child-George-Comstock
/dp/0123725429

[2] Children's Advertising, Proposed Trade Regulation Rulemaking,
43 Fed. Reg. 17967, 17969 (Apr. 25, 1978). See, e.g.,
http://www.law.indiana.edu/fclj/pubs/v58/no2/Ramsey.pdf

[3] From Susan Linn, "Consuming Kids,"
http://topdocumentaryfilms.com/consuming-kids/

[4] Joel Bakan, "The Kids Are Not Alright," NY Times The Opinion
Pages, August 21, 2011.
http://www.nytimes.com/2011/08/22/opinion/corporate-interests-thr
eaten-childrens-welfare.html

[5] Joseph Stiglitz, "Of the 1%, by the 1%, for the 1%," Vanity Fair, May 2011.
http://www.vanityfair.com/society/features/2011/05/top-one-percent-201105

[6] http://en.wikipedia.org/wiki/2011_Egyptian_Revolution

[7] http://en.wikipedia.org/wiki/2011_Indian_anti-corruption_movement

[8] http://occupywallst.org/

[9] http://www.occupytogether.org/

[10] Occupy Wall Street, "Declaration of Occupy Wall Street," Sep. 29, 2011.
http://occupywallst.org/forum/first-official-release-from-occupy-wall-street/

[11] Jean Twenge, "Generation Me: Why Today's Young Americans Are More Confident, Assertive, Entitled--and More Miserable Than Ever Before," Free Press, 2006.
http://www.amazon.com/Generation-Americans-Confident-Assertive-Entitled/dp/0743276973

[12] Thandie Newton, "Embracing Otherness, Embracing Myself," TED talk, July 2011.
http://www.ted.com/talks/thandie_newton_embracing_otherness_embracing_myself.html

[13] Jonathan Haidt, "When Compassion Leads to Sacrilege," Stanford Center for Compassion and Altruism Research, April 2011.
http://ccare.stanford.edu/content/jonathan-haidt-when-compassion-leads-sacrilege

[14] Anthony DeMello, "Awareness: The Perils and Opportunities of Reality," Image Publishers, June 1990.
http://www.amazon.com/Awareness-Opportunities-Reality-Anthony-Mello/dp/0385249373

[15] Kelly McGonigal, "The Power of Self-Compassion," Stanford Cener for Compassion and Altruism Research, Feb. 2011. http://kellymcgonigal.com/2011/08/16/the-power-of-self-compassion/

[16] This statistic is found in Niall Ferguson's TED talk at http://www.ted.com/talks/niall_ferguson_the_6_killer_apps_of_prosperity.html

[17] http://www.tarsandsaction.org/

[18] Quoted in Joe Nocera, "Killing Jobs and Making us Sick," The Opinion Pages, NY Times, Sep. 16, 2011. http://www.nytimes.com/2011/09/17/opinion/nocera-killing-jobs-and-making-us-sick.html

[19] Michael Pollan, "In Defense of Food: An Eater's Manifesto," Penguin, April 2009. http://www.amazon.com/gp/product/0143114964

[20] http://www.millenniumrestaurant.com/

[21] Over 3000 recipes can be found in just http://www.ivu.org/recipes/

[22] Kelly Peloza, "The Vegan Cookie Connoisseur: Over 140 Simply Delicious Recipes That Treat the Eyes and Taste Buds," SkyHorse Publishing, Nov. 2010. http://www.amazon.com/Vegan-Cookie-Connoisseur-Delicious-Recipes/dp/161608121X

[23] http://chefchloe.com/

[24] Tara Parker Pope, "Tasty Vegan Food: Cupcakes Show it Can be Done," NY Times, Sep. 5, 2010. http://well.blogs.nytimes.com/2010/09/06/tasty-vegan-food-cupcakes-show-it-can-be-done/

[25]https://www.facebook.com/group.php?gid=18116069888&v=wall

[26] Simon SInek, "How Great Leaders Inspire Action," Sep. 2009. http://www.ted.com/talks/simon_sinek_how_great_leaders_inspire_action.html

Epilogue:

[1] http://v-dogfood.com/

[2] http://www.irobot.com/

[3] http://en.wikipedia.org/wiki/Dodo

[4]http://www.intel.com/design/network/products/lan/controllers/82540.htm

[5] Scene from "Animals are Beautiful People," 1974. http://www.youtube.com/watch?v=NxkvY2FZRQg

[6] Steve Jobs, "How to Live Before You Die," Commencement Address at Stanford University, Stanford, CA, June 2005. http://www.ted.com/talks/steve_jobs_how_to_live_before_you_die.html

[7] http://www.youtube.com/watch?v=PsdgBwIE5CA

[8] http://pangea.stanford.edu/programs/eiper/

Index

CPSIA information can be obtained
at www.ICGtesting.com
Printed in the USA
LVHW081830090220
646320LV00008B/530